Fluid Mechanics and Its Applications

Founding Editor

René Moreau

Volume 129

Series Editor

André Thess, German Aerospace Center, Institute of Engineering Thermodynamics, Stuttgart, Germany

The purpose of this series is to focus on subjects in which fluid mechanics plays a fundamental role. As well as the more traditional applications of aeronautics, hydraulics, heat and mass transfer etc., books will be published dealing with topics, which are currently in a state of rapid development, such as turbulence, suspensions and multiphase fluids, super and hypersonic flows and numerical modelling techniques. It is a widely held view that it is the interdisciplinary subjects that will receive intense scientific attention, bringing them to the forefront of technological advancement. Fluids have the ability to transport matter and its properties as well as transmit force, therefore fluid mechanics is a subject that is particulary open to cross fertilisation with other sciences and disciplines of engineering. The subject of fluid mechanics will be highly relevant in such domains as chemical, metallurgical, biological and ecological engineering. This series is particularly open to such new multidisciplinary domains. The median level of presentation is the first year graduate student. Some texts are monographs defining the current state of a field; others are accessible to final year undergraduates; but essentially the emphasis is on readability and clarity.

Springer and Professor Thess welcome book ideas from authors. Potential authors who wish to submit a book proposal should contact Dr. Mayra Castro, Senior Editor, Springer Heidelberg, e-mail: mayra.castro@springer.com

Indexed by SCOPUS, EBSCO Discovery Service, OCLC, ProQuest Summon, Google Scholar and SpringerLink

Anders Hedegaard Hansen

Fluid Power Systems

A Lecture Note in Modelling, Analysis and Control

Springer

Anders Hedegaard Hansen
AAU Energy
Aalborg University
Aalborg, Denmark

ISSN 0926-5112 ISSN 2215-0056 (electronic)
Fluid Mechanics and Its Applications
ISBN 978-3-031-15091-3 ISBN 978-3-031-15089-0 (eBook)
https://doi.org/10.1007/978-3-031-15089-0

This Springer imprint is published by the registered company Springer Nature Switzerland AG
The registered company address is: Gewerbestrasse 11, 6330 Cham, Switzerland

Preface

With this lecture note the author strives for an easily accessible note covering some of the fundamental topics in fluid power technology. The main objectives are to enhance the reader's learnings and depth of understandings in the exciting area of fluid power technology—and to achieve these goals, significant space is allocated for detailed derivations of formulas that form the basis of the theory. After successful completion of this note the reader knows how to properly (i) design basic fluid power systems, (ii) construct lumped parameter models of simple fluid power systems, (iii) perform frequency analysis of fluid power components and systems and (iv) develop controllers for fluid power systems. The note mainly focusses on mathematical modelling and analysis of fluid power components and systems, i.e. practical issues such as working principles and construction of components are not treated in details. Neither are system topology selection nor sizing of components using design standards treated in this note.

The text is organised in four main parts that mildly overlap with each other: I Physics of Fluid, II Fluid Power Components, III Fluid Power Systems and IV Learning by Doing.

The topics in Part I Physics of Fluids focus on giving the reader a basic understanding of fluid mechanics related to fluid power technologies. Special focus is placed on modelling of fluid flow through restrictions, viscous forces in fluid flow and how changes in fluid momentum may impact components and systems.

In the Part II Fluid Power Components, the most important and commonly used components are introduced. Emphasis is placed on component modelling and using the derived models the function of each component is illustrated. After completion of Part II, the reader has a toolbox of component knowledge that can be used to put together a system that can perform a certain task.

The third Part III Fluid Power Systems gives an in-depth presentation of system modelling and system analysis followed by a discussion of closed-loop control techniques applicable in fluid power systems. Time domain lumped parameter models are developed for a few example systems. Based on time domain models, linear models are constructed for the purposes of both frequency domain analysis and controller design. Part III concludes with a chapter introducing a "cook book" approach to fluid power system design, which allows the design engineer to derive an initial system design in a systematic way.

The Part IV Learning by Doing is the most important in striving for the objectives of learning and understanding topics of fluid power technology. Part IV is filled with exercises, solution examples and solution strategies. The author truly believes that the knowledge gained from this part of the note is vital for the fluid power technology student: One may read the three prior parts and get an idea of the theories and methods, but for true and profound learning one must acquire the theory by (i) carefully studying of explanatory examples, (ii) independently solve exercises and (iii) become proficient in the art of evaluating the validity and soundness of results.

Aalboro, Denmark
January 2023

Anders Hedegaard Hansen
Associate Professor

Contents

Chapter 1
Introduction—It's Just a Gear

Abstract This chapter introduces fluid power technology and exemplifies three hydrostatic fluid power systems. Each of the three systems shows how fluid power technology is used as a gearing between a mechanical input motion and a mechanical output motion. The chapter ends with a discussion of the units used when studying fluid power systems.

Fluid power technology is a power transmission technology transmitting power by use of a fluid. In a fluid power transmission system a mechanical motion is transformed into a movement of a fluid. Piping and hoses guide the fluid about and transmit large amounts of power. The motion of the fluid is transformed back into a mechanical motion at the proper place

Comparing fluid power technology with electricity, one may transform mechanical motion into electrical power by using a generator. Electrical power is transmitted through copper cables and by using either rotational or linear motors electric power is transformed back into a mechanical motion.

A system transforming one mechanical motion with a given velocity and force to another mechanical motion with a different velocity and force, is often called a gear or a transmission. At a basic level, a fluid power system is just such a transmission system. A fluid power transmission system, however, comes with inherent controlling possibilities, e.g. the "gearing ratio" may easily be controlled, one may build in safety function as overload protection.

1.1 Hydrodynamics Versus Hydrostatics

Fluid transmission technology covers a wide range of technologies with two main branches, hydrodynamics and hydrostatics. In hydrodynamic systems kinetic energy in the fluid is used to drive a mechanical motion whereas in hydrostatic systems the "potential energy" is used to drive the motion.

A. H. Hansen, *Fluid Power Systems*, Fluid Mechanics and Its Applications 129,
https://doi.org/10.1007/978-3-031-15089-0_1

1.1.1 Kinetic Power and Hydrodynamics

In hydrodynamic systems power is transmitted by the kinetic energy in the fluid and the force onto some object is given from change in fluid momentum, with momentum given as:

$$MOMENTUM = MASS \times VELOCITY, \tag{1.1}$$

where if the fluid is to perform work at some object the velocity or mass are required to change.

1.1.2 Pascal's Principle and Hydrostatics

Hydrostatic fluid power transmission technology builds on Pascal's Law, which states:

Pressure in a confined fluid is transmitted undiminished in all directions, and acts with equal force on equal areas, and so at right angles to the container wall.

Pascal's principle states that pressure in a fluid at rest is equal everywhere in the fluid. (Gravitation will however imply a small pressure gradient in the vertical direction.) The force, F, is given as the pressure, p multiple an area, A on which it works:

$$F = Ap. \tag{1.2}$$

The velocity, v, of a fluid flow is given from the volume flow, Q, and the flow area A as:

$$v = \frac{Q}{A}. \tag{1.3}$$

With velocity and force given power may be given as:

$$\mathcal{P} = Force \times Velocity, \tag{1.4}$$

such that power is given as:

$$\mathcal{P} = Ap \times \frac{Q}{A}, \tag{1.5}$$

and may be reduced to

$$\mathcal{P} = Pressure \times Volumeflow. \tag{1.6}$$

1.1.3 *Three Hydrostatic Systems*

Let's introduce three simple fluid power systems based on Pascal's Principle and such hydrostatics.

Fluid power Lever power Jack

In a fluid power lever operated by hand, one translational motion is transformed into an other translational motion. The function may be illustrated clearly with the system shown in Fig. 1.1 where a small force over a long distance is converted to a larger force over a short distance. In the fluid power lever, the fact of Pascal's principle of equal pressure everywhere in the container is utilised.

The hydrostatic force on the two pistons is given as:

$$F_1 = pA_1, \tag{1.7}$$

$$F_2 = pA_2, \tag{1.8}$$

where p is the pressure in the fluid and A_1 and A_2 are the areas of the two pistons. Let a worker apply force, F_1, on the small piston, then the resulting force on the large piston may simply be found by:

$$p = \frac{F_1}{A_1}, \tag{1.9}$$

$$F_2 = pA_2 = \frac{F_1}{A_1}A_2 = \frac{A_2}{A_1}F_1, \tag{1.10}$$

when one uses Pascal's law to assume equal pressure in piston 1 and 2 (pressure difference due to gravitation is neglected). The force exerted by the worker is multiplied with the area ratio to yield the force exerted by piston 2. The fluid power system is merely a gearing that transmits power.

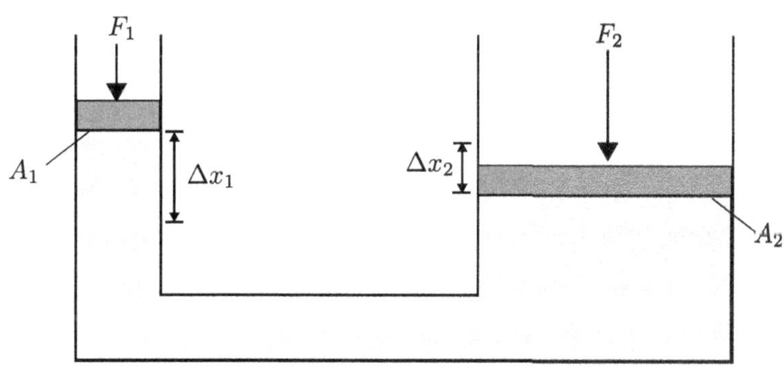

Fig. 1.1 Illustration of a simple fluid power system—a fluid power lever

The gearing in tells that a large movement at low force is transformed to a small movement at high force. Let's assume the system in Fig. 1.1 to be loss free, then the work by force F_1 and F_2 must be equal such as:

$$\mathcal{W}_1 = F_1 \Delta x_1, \tag{1.11}$$

$$\mathcal{W}_2 = F_2 \Delta x_2, \tag{1.12}$$

$$F_1 \Delta x_1 = F_2 \Delta x_2 = \tfrac{A_2}{A_1} F_1 \Delta x_2, \tag{1.13}$$

$$\Delta x_1 = \tfrac{A_2}{A_1} \Delta x_2, \tag{1.14}$$

The area ratio is again seen to be a gearing ratio between position change of piston 1 and 2 respectively. One may imagine the movement to take place during time δt and one will get the relation between velocities.

1.1.3.1 Rotational to Translational Movement

Let's briefly introduce a rotational to translational gear. In Fig. 1.2 what we shall learn to be a vane pump is driven by a rotational torque τ at a velocity n producing a volume flow Q of fluid into the "container".

Let's assume the volume flow from the pump to be proportional to the rotational velocity, such $Q = Dn$, where D is a proportionality constant we will study later. One may now calculate the velocity of the piston as the ratio between volume flow and piston area:

$$\dot{x}_1 = \frac{Q}{A}. \tag{1.15}$$

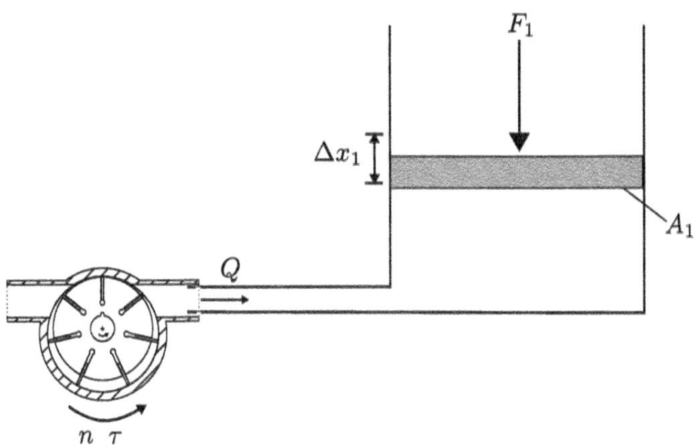

Fig. 1.2 Illustration of a simple fluid power system

With volume flow in the unit of cubic meter per seconds, and area in the unit square meter, one may be convinced of the validity of the formula. The gearing i_{gear} from rotational velocity to translational velocity is given as:

$$i_{\text{gear}} = \frac{D}{A}, \tag{1.16}$$

and the velocity of the piston is given as:

$$\dot{x}_1 = \frac{D}{A}n. \tag{1.17}$$

1.1.3.2 Rotational to Rotational Movement

Lastly we consider a fluid power transmission system that has a rotational motion as input and a rotational motion as output. In Fig. 1.3 we again see a vane pump, however now driving a vane motor.

Using D_1 and D_2 for the proportionality constants of the pump and motor respectively one gets:

$$Q = D_1 n_1, \tag{1.18}$$

$$n_2 = \frac{Q}{D_2} = \frac{D_1}{D_2} n_1, \tag{1.19}$$

why the gearing between input and output velocity becomes $\frac{D_1}{D_2}$. Assuming the same two proportionality constants for the pressure and torque relation we get:

$$\tau_1 = D_1 p, \tag{1.20}$$

$$\tau_2 = D_2 p = \frac{D_2}{D_1}\tau_1 \tag{1.21}$$

Pascals Principle dictates that the pressure in the volume between the two units is constant. Naturally torques are geared with the same ratio as velocities, however opposite.

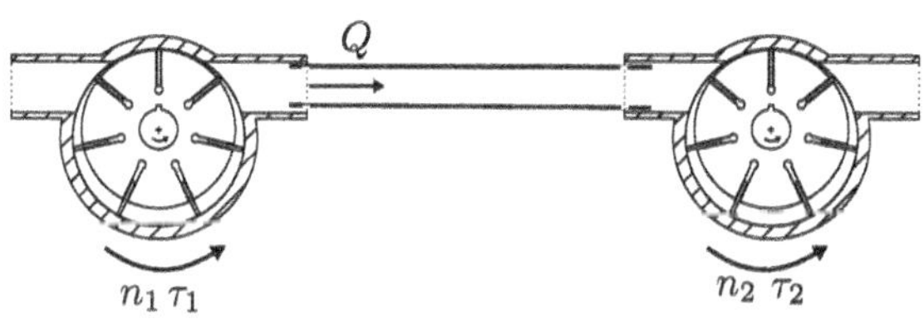

Fig. 1.3 Illustration of a simple fluid power system

1.2 Hydraulic Systems Versus Fluid Power Systems

In much literature the term hydraulics is used for what in this note is named fluid power systems, even though it is exactly the same technology being studied. To avoid confusion this note uses "fluid power" rather than "hydraulic" because both personal conversations and professional corporations have been haunted by misunderstandings between engineers with a Mechanical Engineering background and engineers with a Civil Engineering background. One thinks of ground water modelling, ocean wave loadings on off-shore structure; while another is thinking of orifice flow in valves and pressure forces on piston heads in cylinders. Therefore fluid power systems is used rather than hydraulic systems, yet they are the same.

1.3 Units in Fluid Power Systems

In fluid power technology there is a history of not using SI-units as the preferred unit system. Even though SI-Units are gaining popularity, litres per minute and bar are still most often used for volume flow and pressure respectively. Even more confusingly, there is also a difference in preferred units based on geography, e.g. gallons per minute in USA etc.

In this note we distinguish between SI-Units and HYD-Units as we shall name them. Table 1.1 gives an overview of the units used for various parameters—some we know and some we shall familiarise ourselves with in studying fluid power components and systems through this note.

Table 1.1 Overview of SI- and HYD-Units used in fluid power systems

	Symbol	SI-units	HYD-unit	Conversion
Pressure	p	Pa—Pascal (N/m^2)	bar	10^5 Pa = 1 bar
Volume flow	Q	m^3/s	L/min	$1\ m^3/s = 60000$ L/min
Pump displacement	D_p	m^3/rad	ccm/rev	$1\ m^3/rad = 2\pi\, 10^6$ ccm/rev
Motor displacement	D_m	m^3/rad	ccm/rev	$1\ m^3/rad = 2\pi\, 10^6$ ccm/rev
Velocity	v	m/s	m/s	
Volume	V	m^3	L	$1\ m^3 = 1000$ L L
Dynamic viscosity	μ	Ns/m^2	P (Poise)	$1\ Ns/m^2 = 10$ P
Kinematic viscosity	ν	m^2/s	cSt (Centistokes)	$1\ m^2/s = 10^6$ cSt

Even though industry and technicians commonly use HYD-Units, it is recommended to only use SI-Units when performing calculations and analyses, though Hyd-Units may be used when presenting results.

Part I
Physics of Fluid

In Fluid Power Technology, power is transmitted using a fluid. Naturally the utilised fluid influences system behaviour, both in terms of performance and ware. In fluid power systems oil is the preferred fluid even though other fluids e.g. water may be used. The preferred oil types range from mineral oils to synthetically produced oils. With a focus on environmental issues, biodegradable oil types are developed and introduced where oil spill may damage the environment (e.g. offshore installation).

With power hydro-statically transmitted, the medium in the form of the fluid is important. With regard to system performance viscosity, compressibility and density is especially important. Whereas lubrication ability, flammability, foam-ability and degrading are important parameters with regard to system ware and piratical design issues.

In this lecture note, the focus is on the performance and functionality of the fluid power systems where we investigate the fluid parameters; viscosity, compressibility and density. Furthermore, fluid kinematics and dynamics are investigated in relation to fluid power systems. This part of the lecture note is not designed to substitute a proper textbook on fluid mechanics, but is merely an introduction, preparing the reader for the study of fluid behaviour in fluid power systems. For further knowledge on fluid mechanics the reader should refer to a dedicated textbook as. e.g. [1].

Chapter 2
Fluid Parameters

Abstract This chapter introduces viscosity, density, and compressibility, as these are important fluid parameters when dealing with fluid power systems. Firstly, some simple viscosity models are introduced, this is both pressure- and temperature-dependent models. The viscous force from a fluid on to a moving piston is investigated using the definition of dynamic viscosity. Secondly, density and compressibility are introduced in parallel. Most important for fluid power systems is the introduction of models for bulk modulus and density of a fluid-air mixture, where significant changes in bulk modulus is seen at low pressure levels.

Viscosity, density and compressibility are important parameters of the fluid in fluid power systems. The topics of this chapter are: an introduction to viscosity models and shear forces in Newtonian fluids, a walk-through of the derivation of models for density, and bulk modulus of a fluid-air mixture.

Initially the topics viscosity and fluid density and compressibility are introduced followed by reflections on how these topics relate to fluid power system. This lecture note is solely based on Newtonian fluids.

2.1 Viscosity

The dynamic viscosity relates to motion and shearing stress in a fluid flow. That is why the dynamic viscosity is referred to as the shearing viscosity. Let us use the well known set-up for studying the relation between shearing stresses and fluid velocity. The set-up consisting of a fluid in-between two parallel plates is illustrated in Fig. 2.1.

The plate named 1 is stationary, i.e. has zero velocity in the general reference frame, while plate 2 is moving with a constant positive velocity in direction x. Assuming no slip conditions on the surface of the plates yield that the fluid layer touching plate 1 will have zero velocity, while the fluid layer touching plate 2 will have the x-direction velocity $\dot{x}_2$ that plate 2 has. The velocity of a fluid layer between the two plates will increase linearly from 0 at the surface of plate 1 to $\dot{x}_2$ at the surface of plate 2. This is also illustrated by the velocity field indicated by arrows in Fig. 2.1.

A. H. Hansen, *Fluid Power Systems*, Fluid Mechanics and Its Applications 129,
https://doi.org/10.1007/978-3-031-15089-0_2

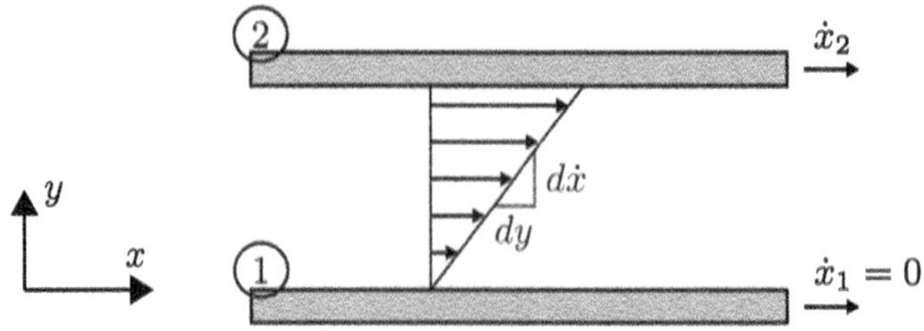

Fig. 2.1 Illustration of the velocity field of a Newtonian fluid between two plates

The velocity of the fluid is induced by shearing stresses initially from the surface of plate 2 onto the fluid and subsequently by shearing stresses between the fluid layers. Shear stress between adjacent fluid layers in Newtonian fluids is given as:

$$\tau = \mu \frac{\mathrm{d}\dot{x}}{\mathrm{d}y} . \tag{2.1}$$

The shear stress in the fluid is proportional to the ratio of change in velocity, $\mathrm{d}\dot{x}$, to the change in position, $\mathrm{d}y$, perpendicular to the flow direction. The proportionality constant is the dynamic viscosity which is given as:

$$\mu = \tau \frac{\mathrm{d}y}{\mathrm{d}\dot{x}} . \tag{2.2}$$

In addition to the dynamic viscosity the kinematic viscosity is often used. The kinematic viscosity is defined as the ratio of dynamic viscosity to density. The symbol ν is, in this note, used for the kinematic viscosity which is defined as:

$$\nu = \frac{\mu}{\rho} , \tag{2.3}$$

where ρ is fluid density.

Investigating the unit for viscosity one notes that the dynamic viscosity is (force $\times$ time/area) with SI-unit Ns/m^2 while the kinematic viscosity is (length^2/time) with SI-unit m^2/s. As introduced in Sect. 1.3 data sheets from industry often use HYD-Units, which in case of viscosity is often Poise or Stokes, see Table 2.1. The fluid parameter viscosity depends on the actual fluid used and the state of said fluid.

Table 2.1 HYD units for viscosity

Name	Symbol	SI units
Poise	1 P	$0.1\ Ns/m^2$
Centistokes	1 cSt	$10^{-}6\ m^2/s$

Viscosity is especially dependent on fluid temperature and pressure. In the following section a couple of viscosity models are introduced, some temperature dependent and some both temperature and pressure dependent.

2.1.1 Viscosity Models

The viscosity of fluids change both with pressure and temperature. Most profound is the change with temperature. In fluid power systems the design engineer often aims at obtaining a constant operational temperature of the fluid, however this may not always be possible, for example in mobile applications working in various environments e.g. −30–30 °C, for a forestry machine in northern Europe, or in the varying conditions inside the nacelle of an off-shore wind turbine.

As we shall see later the viscosity is often modeled as constant in fluid power systems, however some viscosity models are briefly introduced and may be required in some cases. In Table 2.2 five viscosity models are given, the first two depending on temperature, the third depending on pressure while the last two are depending on both temperature and pressure.

For all the models given in the table T and p are temperature and pressure of the fluid respectively, other parameters are constants that may vary from oil to oil. In the paper [1] the parameters for the Vogel model have been measured, a plot of the viscosity as a function of temperature may such be given in Fig. 2.2.

The Barus model includes pressure changes and gives the viscosity as a function of pressure change, however the temperature change is only included in the pressure-viscosity coefficient $\alpha(T)$. μ_0 may though be chosen for various temperatures. The Barus model takes a viscosity at atmospheric pressure μ_0 and multiplies this with a pressure varying term which is shown in Fig. 2.3. The constants found in [1] are once again used for illustration.

By combining Barus' and Vogel's models, a viscosity model depending on both temperature and pressure is obtained. Employing the parameters found in [1] the viscosity curves seen in Fig. 2.4 appear. It is evident that temperature variation influ-

Table 2.2 Viscosity models depending on temperature and/or pressure

Name	Model
Reynolds	$\mu_0(T) = b_R \exp(-K_\mathrm{T}(T))$
Vogel	$\mu_0(T) = a \exp\left(\frac{b}{T-c}\right)$
Barus	$\mu(T, p) = \mu_0(T) \exp(\alpha(T)p)$
Barus-Reynolds	$\mu(T, p) = b_R \exp(\alpha(T)p - K_\mathrm{T}(T))$
Barus-Vogel	$\mu(T, p) = a \exp\left(\frac{b}{T-c}\right) \exp(\alpha(T)p)$
	$\alpha(T) = \frac{1}{a_1+a_2T+(b_1+b_2T)p}$

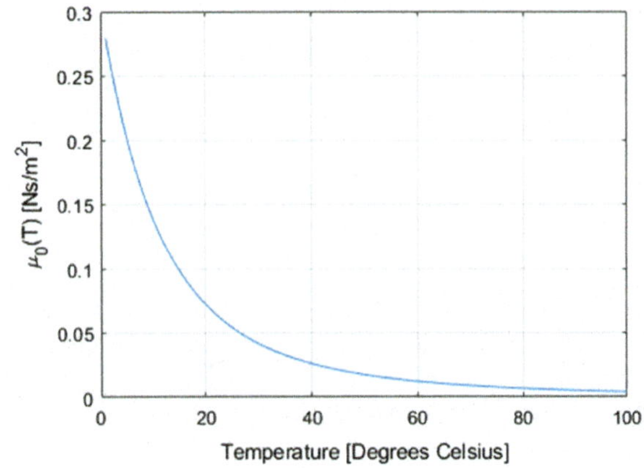

Fig. 2.2 Illustration of Vogel viscosity model. For HM32 $\Rightarrow\ a = 0.0000736317$; $b = 797.7122$; $c = 117.3562$ [1]

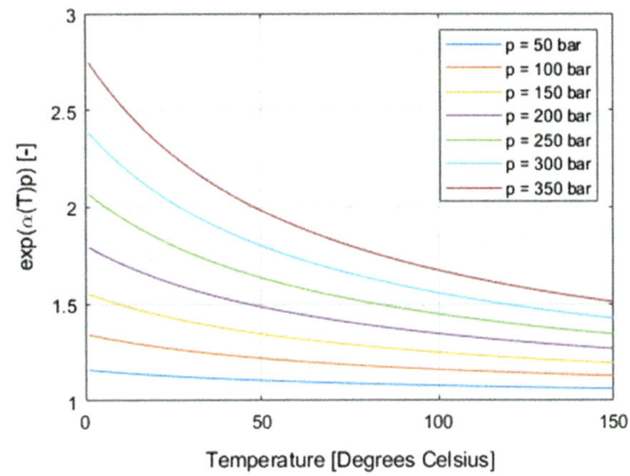

Fig. 2.3 Illustration of Barus model

ences the viscosity significantly more than pressure variation, when using common values for fluid power systems. ($T = [30\ 80]^\circ$ C and $p = [0\ 350]$ bar).

However, if the design engineer succeeds in keeping a constant oil temperature, the pressure dependent part may become of importance. For more complex models and further information on viscosity models the reader should refer to other materials.

2.1.2 *Viscous Force Due to Fluid Flow*

A viscous fluid flowing through or passing an object will exert a force on that object. A part of the force from the fluid onto the object will be due to shear stresses induced by the viscous fluid. This section gives two examples of forces induced by a viscous fluid flow. Note that force exerted by a fluid includes both pressure force (normal stresses) and shear stresses, however the examples given next focus on the latter.

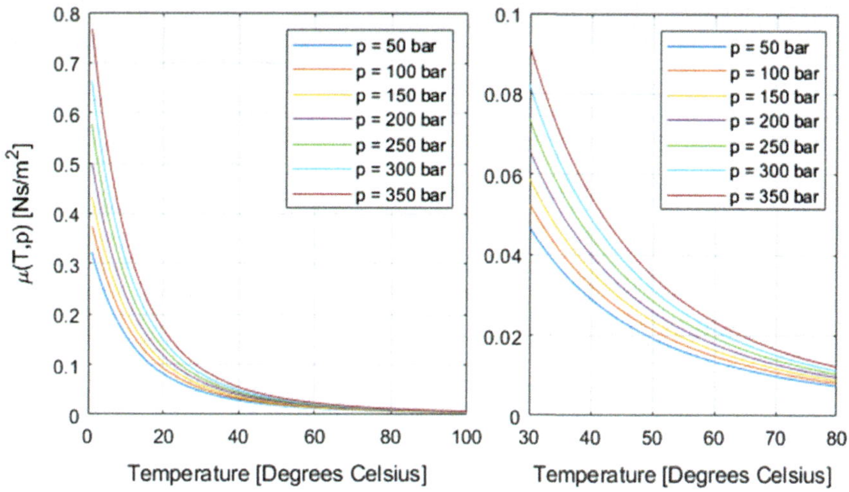

Fig. 2.4 Illustration of Barus-Vogel viscosity model, with zoom shown right for common operating temperatures

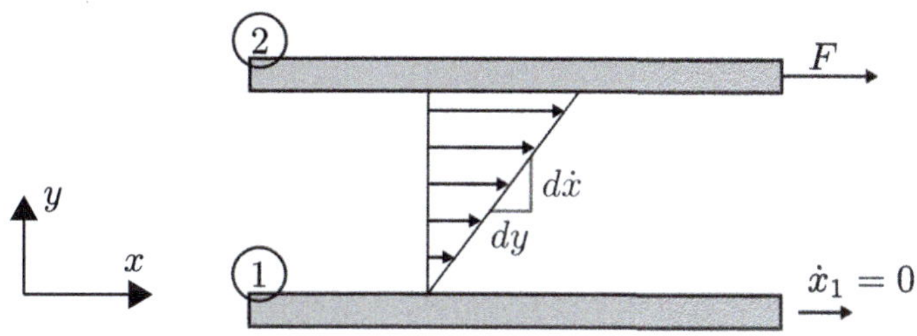

Fig. 2.5 Illustration of the velocity field of a Newtonian fluid between two plates when applying a force on plate two

Viscous Force on Parallel Plates

When two parallel plates move relative to each other with a fluid film in between, a shear force is exerted on the plates. Let us revisit the setup with the two parallel plates. In Fig. 2.5 plate one is again fixed to the global reference frame and stationary while we exert a force, F on plate 2 in the positive x-direction.

The force F is counteracted by a force from the fluid which is given by the shear stress multiplied the surface area. The shear stress is:

$$\tau = \frac{F}{A}, \tag{2.4}$$

where A is the surface area of the plate touching the fluid. Using (2.1) the shear stress is:

$$\mu \frac{\mathrm{d}\dot{x}}{\mathrm{d}y} = \frac{F}{A}. \tag{2.5}$$

Now let us set the distance between the two plates to h and the velocity of plate 2 is $\dot{x}_2$, when the force requested for that given velocity is:

$$F = \frac{\mu A}{h}\dot{x}_2. \tag{2.6}$$

The force required to uphold the steady state velocity $\dot{x}_2$ is in direct proportion to the viscosity multiple by the surface area, while inversely proportional to the distance between the two parallel plates.

Viscous Force on a Piston

A circular piston moving in a stationary tube is illustrated in Fig. 2.6. Such a case is found in e.g.; fluid power cylinders where a piston moves in a cylinder barrel, or valves where the valve spool or plungers move in a valve hosing. In the case study we assume zero pressure on the left and right hand side of the piston. Forces on the piston are such solely the viscous forces.

Let us assume that the clearance between the tube and the piston is very small compared to the diameter of the piston. One can approximate the annular flow area between the piston and tube as the flow area between two parallel plates and one can utilise the definition of shear stress in a Newtonian fluid:

$$\tau = \mu \frac{d\dot{x}}{dy}. \tag{2.7}$$

By separation of variables and integration, the stress at the surface is found.

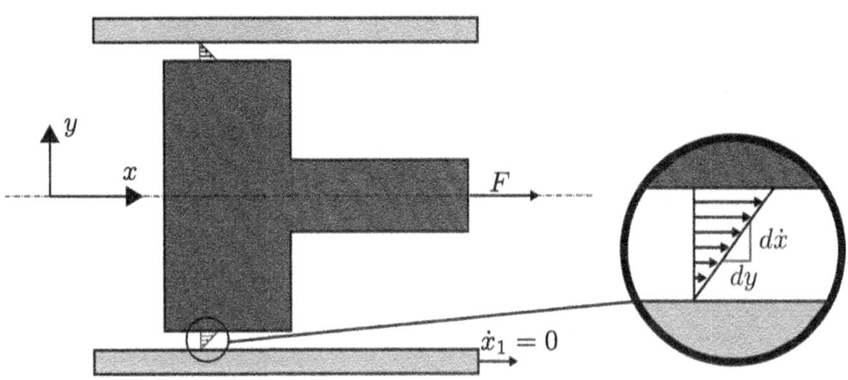

Fig. 2.6 Illustration of viscous force on a piston moving in a cylinder barrel

$$\tau dy = \mu d\dot{x} \tag{2.8}$$

$$\tau \int dy = \mu \int d\dot{x} \tag{2.9}$$

$$\tau = \mu \frac{\dot{x}_f - \dot{x}_i}{y_f - y_i}, \tag{2.10}$$

where $\dot{x}_f$, $\dot{x}_i$, y_f, y_i are the integration limits as initial and final velocity at initial and final position respectively. The shear stress at the surface is known and the shear force is given as,

$$F = A_s \tau_s, \tag{2.11}$$

$$F = D\pi L \mu \frac{\Delta \dot{x}}{\Delta y}. \tag{2.12}$$

The shear force is again seen as proportional to the relative velocity over the gap height. The proportionality constant is here the fluid viscosity, μ multiplied by the surface area, $D\pi L$.

Recapitulating, we found that the shear force acting on a piston moving in a tube is proportional to the relative velocity of the piston and tube divided by the gap height between the piston and the tube. One should note that we assume that only the shearing force in the fluid in the gap is working on the piston. As evident from Newton's third law the force required to move the piston is equal to the force required to hold the tube, such that it is not moved by the shearing force induced in the fluid due to the piston movement.

2.2 Fluid Density and Compressibility

Fluid density and stiffness are two important parameters when modeling fluid power systems. The current section features a brief introduction to some models for fluid stiffness and density. Further and more important for real life fluid power systems, fluid stiffness and density of a fluid air mixture is investigated.

2.2.1 *Equation of State for a Fluid*

With the density of the working fluid named ρ the variation of density may be modeled with a first order Taylor expansion. With ρ_0 denoting the density at a given initial point, the Taylor expansion in pressure and temperature yields:

$$\rho = \rho_0 + \left.\frac{\partial \rho}{\partial p}\right|_{\mathrm{T}} (p - p_0) + \left.\frac{\partial \rho}{\partial T}\right|_{\mathrm{p}} (T - T_0) + \mathcal{O}^2. \tag{2.13}$$

Defining two constants β and α including the partial derivatives of the density function with respect to temperature and pressure respectively, the equation reduces to

$$\rho = \rho_0 \left[1 + \frac{1}{\beta}(p - p_0) - \alpha(T - T_0)\right] + \mathcal{O}^2, \tag{2.14}$$

where the bulk modulus β is defined as

$$\beta \equiv \rho_0 \left.\frac{\partial p}{\partial \rho}\right|_T, \tag{2.15}$$

and where:

$$\alpha \equiv -\frac{1}{\rho_0} \left.\frac{\partial \rho}{\partial T}\right|_p. \tag{2.16}$$

Note that the constants β and α include the density ρ_0 at initial condition and the partial derivative of the density function with respect to temperature and pressure respectively. Bulk modulus, β, tells the relation between density change and pressure change when the temperature is held constant. Likewise the cubical expansion coefficient, α, tells the relation between density change and temperature change when the pressure is held constant.

With conservation of mass, the mass at time zero $m = \rho_0 V_0$ is the same as the mass at time 1 $m = \rho_1 V_1$ which leads to the bulk modulus in terms of initial volume and volume change rather than initial density and density change:

$$\beta \equiv -V_0 \left.\frac{\partial p}{\partial V}\right|_T. \tag{2.17}$$

Isolating the pressure change from the definition of bulk modulus, it is given that:

$$dp = -\frac{\beta}{V_0} dV, \qquad dp = \frac{\beta}{\rho_0} d\rho. \tag{2.18}$$

2.2.2 Pressure Dependent Density and Bulk Modulus of Fluid-Air Mixture

In fluid power systems the fluid used is never pure, some air will always be present in the fluid. Both density and bulk modulus are very different for air and oil. Non-dissolved air-bubbles play a significant role when performing dynamic modelling of fluid power systems, especially at low pressure levels (0–50 bar). Assuming the working fluid being a uniform fluid leads to the request for an effective density and bulk modulus for the fluid-air mixture making up the "uniform" fluid.

Let V_0 be the total volume in Fig. 2.7 and let α be the ratio of air volume to the total volume at atmospheric pressure. Then the air and fluid volume at atmospheric pressure are given as:

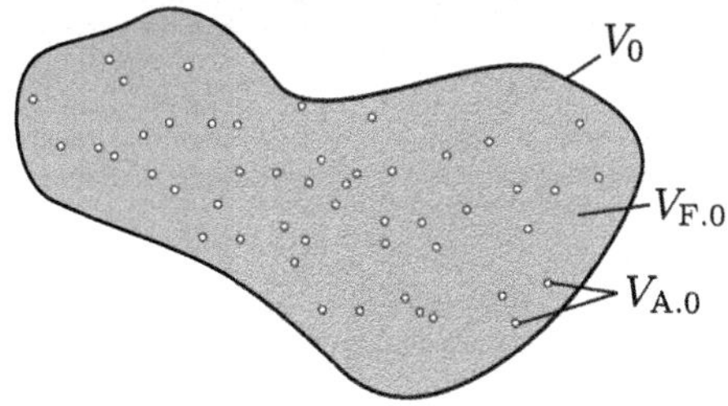

Fig. 2.7 Illustration of air-fluid mixture, with total volume V_0, fluid volume $V_{F,0}$ and air volume $V_{A,0}$

$$V_{A,0} = \alpha V_0, \tag{2.19}$$

$$V_{F,0} = (1 - \alpha) V_0. \tag{2.20}$$

With an increase in pressure from p_0 to p the fluid-air mixture will be compressed, resulting in a change in the effective density and bulk modulus of the mixture. This effect is the topic of the following subsections where the compression of the pure fluid and pure air is investigated and combined to effective values for the mixture.

One should note that pressure in the fluid-air mixture is equal throughout the control volume.

Density

When dealing with a constant amount of mass and respecting conservation of mass, one may find the effective density:

$$\rho = \frac{\text{mass}}{\text{volume}}, \tag{2.21}$$

$$\rho_{\text{eff}} = \frac{\rho_{A,0} V_{A,0} + \rho_{F,0} V_{F,0}}{V_A + V_F}, \tag{2.22}$$

where $\rho_{A,0}$ and $\rho_{F,0}$ are the air and fluid density at atmospheric pressure, while V_A and V_F are the air and fluid volume at the pressure p. The mass of the matter is found at atmospheric pressure in the nominator and the volume at a given pressure p is given in the denominator. ρ_{eff} is the effective density of the fluid-air mixture when treated as an uniform fluid.

Bulk Modulus

Expressing a confined amount of mass in terms of density and volume yields:

$$\text{mass} = \rho V. \tag{2.23}$$

The time derivative of the mass will be zero as a confined amount of mass is investigated:

$$\frac{d\text{mass}}{dt} = \frac{d}{dt}(\rho V) = \frac{d\rho}{dt} V + \frac{dV}{dt} \rho = 0, \tag{2.24}$$

$$\frac{d\rho}{dt} V = -\frac{dV}{dt} \rho, \tag{2.25}$$

$$\frac{1}{\rho} d\rho = -\frac{1}{V} dV. \tag{2.26}$$

Utilising the definition of bulk modulus in (2.15) yields:

$$\frac{1}{\beta}dp = -\tfrac{1}{V}dV, \tag{2.27}$$

$$\beta = -V\tfrac{dp}{dV}. \tag{2.28}$$

Inserting volume and change in volume for the air-fluid mixture yields:

$$\beta_{\text{eff}} = -\left(V_{\text{F}} + V_{\text{A}}\right)\tfrac{dp}{d(V_{\text{F}}+V_{\text{A}})}, \tag{2.29}$$

$$\beta_{\text{eff}} = -\tfrac{(V_{\text{F}}+V_{\text{A}})}{\frac{dV_{\text{F}}}{dp}+\frac{dV_{\text{A}}}{dp}}. \tag{2.30}$$

We see that the effective density (2.22) and bulk modulus (2.30) of a fluid-air mixture at a given pressure is dependent on the volume ratio of fluid and air respectively. Two functions describing the volume ratio of fluid and air respectively are defined. The function gives the ratio of volume at the pressure p to the fluid volume at the pressure p_0, hence:

$$f_{\text{F}} = \tfrac{V_{\text{F}}}{V_{\text{F},0}} \quad \Rightarrow \quad V_{\text{F}} = V_{\text{F},0}f_{\text{F}} = (1-\alpha)V_0 f_{\text{F}}, \tag{2.31}$$

$$f_{\text{A}} = \tfrac{V_{\text{A}}}{V_{\text{A},0}} \quad \Rightarrow \quad V_{\text{A}} = V_{\text{A},0}f_{\text{A}} = \alpha V_0 f_{\text{A}}. \tag{2.32}$$

Volume Ratio of Air

Assuming the compression of the air part of the fluid-air mixture to be poly-tropic gives:

$$pV^n = \text{constant}, \tag{2.33}$$

$$pV_A^n = p_0 V_{A,0}^n. \tag{2.34}$$

Hence, the volume ratio function is,

$$f_{\text{A}} = \frac{V_{\text{A}}}{V_{\text{A},0}} = \left(\frac{p_0}{p}\right)^{\frac{1}{n}}, \tag{2.35}$$

and the pressure dependent change is,

$$\frac{df_{\text{A}}}{dp} = -\frac{1}{n}\left(\frac{p_0}{p}\right)^{\frac{1}{n}-1}\frac{p_0}{p^2} = -\frac{1}{np_0}\left(\frac{p_0}{p}\right)^{\frac{n+1}{n}}. \tag{2.36}$$

The polytropic index n is a number ranging from 1 for an isothermal process to $\gamma_{\text{air}} = 1.4$ for an adiabatic process. $n = \gamma_{\text{air}}$ is commonly used for air-bubble compression in fluid power systems, as the large pressure gradients leaves limited time for heat-exchange during the compression process.

Volume Ratio of Fluid

Let bulk modulus of the oil be constant, $\beta_F = \beta_0$, and inserting this in the definition of bulk modulus in (2.17) yields,

$$-\frac{1}{V}dV = \frac{1}{\beta_0}dp, \tag{2.37}$$

using the initial fluid volume and pressure $V_{F,0}$ and p_0 as integration limits together with the actual fluid volume, V_F at a given pressure p the volume ratio function may be derived as,

$$\int_{V_{F,0}}^{V_F} -\tfrac{1}{V}dV = \int_{p_0}^{p} \tfrac{1}{\beta_0}dp, \tag{2.38}$$

$$\ln\left(\tfrac{V_F}{V_{F,0}}\right) = -\tfrac{1}{\beta_0}(p - p_0), \tag{2.39}$$

$$f_F = \tfrac{V_F}{V_{F,0}} = \exp\left(\tfrac{p_0 - p}{\beta_0}\right), \tag{2.40}$$

$$\tfrac{df_F}{dp} = -\tfrac{1}{\beta_0}\tfrac{V_F}{V_{F,0}} = -\tfrac{1}{\beta_0}\exp\left(\tfrac{p_0 - p}{\beta_0}\right). \tag{2.41}$$

With bulk modulus of the oil assumed depending on the pressure, such as $\beta_F = \beta_0 + m(p - p_0)$ the volume ratio function is derived as,

$$\int_{V_{F,0}}^{V_F} -\tfrac{1}{V}dV = \int_{p_0}^{p} \tfrac{1}{\beta_0 + m(p - p_0)}dp, \tag{2.42}$$

$$-\ln\left[V\right]_{V_{F,0}}^{V_F} = \tfrac{1}{m}\left[\ln\left(\beta_0 + m(p - p_0)\right)\right]_{p_0}^{p}, \tag{2.43}$$

$$\ln\left[\tfrac{V_F}{V_{F,0}}\right] = -\tfrac{1}{m}\left[\ln\tfrac{\beta_0 + m(p - p_0)}{\beta_0}\right], \tag{2.44}$$

$$f_F = \tfrac{V_F}{V_{F,0}} = \left(1 + \tfrac{m(p - p_0)}{\beta_0}\right)^{-\frac{1}{m}}, \tag{2.45}$$

$$\tfrac{df_F}{dp} = -\tfrac{1}{\beta_0}\left(1 + \tfrac{m(p - p_0)}{\beta_0}\right)^{-\frac{m+1}{m}}. \tag{2.46}$$

We have now derived the volume ratio functions f_F and f_A for the fluid and the air respectively, as well as the change in volume ratio with change in pressure $\frac{df_F}{dp}$ and $\frac{df_A}{dp}$ for the fluid and the air respectively. With these functions known, one may insert in (2.30) and (2.22) set-up the equations for effective density and bulk modulus of the fluid-air mixture.

2.2.3 Summery

Models for the effective density and bulk modulus of a fluid-air mixture are given in the following two boxes. Note that two models for the fluid bulk modulus are used, one with constant fluid bulk modulus and one with a fluid bulk modulus increasing proportionally with the pressure. Note that other models for the separate air and fluid behaviour will yield other models for the uniform fluid model of the fluid-air mixture.

Density of Fluid-Air mixture;
with constant fluid bulk modulus $\beta_F = \beta_0$

$$\rho_{eff} = \frac{\rho_{F,0}(1-\alpha)+\rho_{A,0}\alpha}{(1-\alpha)\exp\left(\frac{p_0-p}{\beta_0}\right)+\alpha\left(\frac{p_0}{p}\right)^{\frac{1}{n}}}, \tag{2.47}$$

with variable fluid bulk modulus $\beta_F = \beta_0 + m(p - p_0)$

$$\rho_{eff} = \frac{\rho_{F,0}(1-\alpha)+\rho_{A,0}\alpha}{(1-\alpha)\left(1+\frac{m(p-p_0)}{\beta_0}\right)^{-\frac{1}{m}}+\alpha\left(\frac{p_0}{p}\right)^{\frac{1}{n}}}. \tag{2.48}$$

Bulk modulus of Fluid-Air mixture;
with constant fluid bulk modulus $\beta_F = \beta_0$

$$\beta_{eff} = \frac{(1-\alpha)\exp\left(\frac{p_0-p}{\beta_0}\right)+\alpha\left(\frac{p_0}{p}\right)^{\frac{1}{n}}}{\frac{1-\alpha}{\beta_0}\exp\left(\frac{p_0-p}{\beta_0}\right)+\frac{\alpha}{np_0}\left(\frac{p_0}{p}\right)^{\frac{n+1}{n}}}, \tag{2.49}$$

with variable fluid bulk modulus $\beta_F = \beta_0 + m(p - p_0)$

$$\beta_{eff} = \frac{(1-\alpha)\left(1+\frac{m(p-p_0)}{\beta_0}\right)^{-\frac{1}{m}}+\alpha\left(\frac{p_0}{p}\right)^{\frac{1}{n}}}{\frac{1-\alpha}{\beta_0}\left(1+\frac{m(p-p_0)}{\beta_0}\right)^{-\frac{m+1}{m}}+\frac{\alpha}{np_0}\left(\frac{p_0}{p}\right)^{\frac{n+1}{n}}}. \tag{2.50}$$

Figure 2.8 shows the effective bulk modulus and density of a fluid-air mixture as a function of the current pressure in the mixture. A constant fluid bulk modulus is used which leads to the effective bulk modulus approaching $\beta_0 = 16000$ bar as the pressure increases. Likewise the slope of the density change approaches a constant value as the pressure increases.

Form the bulk modulus curves it is evident that the variation is significant at low pressure levels and less profound at high pressure levels. This has led engineers to design systems with a relatively high back pressure ensuring stiffness in the system. In later chapters we shall experience the benefit of a relatively high back pressure yielding "high" bulk modulus.

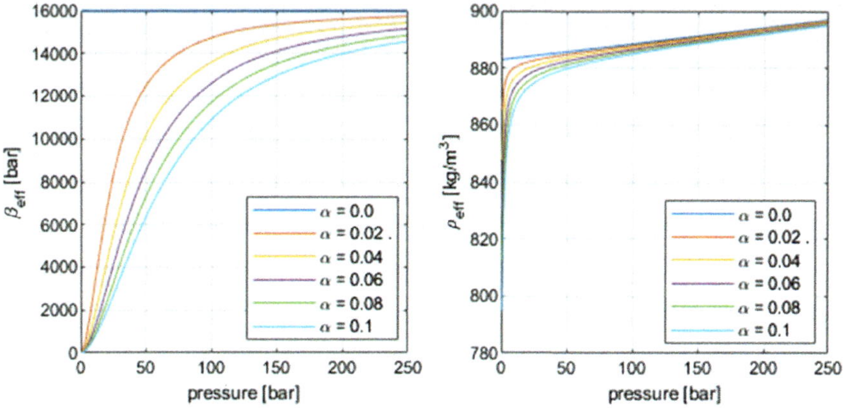

Fig. 2.8 Curves for the effective bulk modulus and density with; $\beta_F = \beta_0 = 16000$ bar; $p_0 = 1$ bar; $\rho_{F,0} = 883$ kg/m^3; $\rho_{A,0} = 1.225$ kg/m^3; $n = 1.4$

Reference

1. Knezevic D Savic V (2006) Mathematical modeling of changing of dynamic viscosity, as a function of temperature and pressure, of mineral oils for hydraulic systems. Mechanical Engineering, vol 4, Issue 1 (2006), pp 27–34

Chapter 3
Fluids Mechanics

Abstract This chapter briefly familiarizes the reader with some topics of fluid mechanics regarding fluid power systems. First, the flow continuity equation is derived from conservation of mass. Secondly, the momentum change of a moving fluid and its influence on fluid power systems is investigated. Lastly, Navier-Stokes equations are introduced from Newtons second law.

The next chapter briefly introduces some basic topics within fluid mechanics that are important for understanding fluid power systems. Both differential and control volume approaches will be discussed. We shall investigate how conservation of momentum and mass may be introduced in fluid systems and how this may affect fluid power components and systems. The main topics will be the flow continuity equation, forces on components due to momentum change in a fluid flow, Navier-Stokes equation and viscous fluid flow. Note, that this chapter is a recap of fluid mechanics in relation of fluid power system and note a proper textbook of fluid mechanics, for that consult e.g., [1] which this chapter leans on for knowledge.

3.1 Conservation of Mass

With mass being neither created or destroyed, it is said to be conserved. Conservation of mass dictates that the mass of a system remains constant as the system moves through a flow field. The change of mass as such is zero, which is mathematically formulated as:

$$\frac{DM_s}{Dt} = 0. \tag{3.1}$$

We shall now use conservation of mass as the foundation in deriving the flow continuity equation. Initially the control volume approach is exploited, followed by the differential volume form.

A. H. Hansen, *Fluid Power Systems*, Fluid Mechanics and Its Applications 129,
https://doi.org/10.1007/978-3-031-15089-0_3

3.1.1 Control Volume Approach

In the control volume approach Reynolds Transport Theorem (RTT) becomes useful, as a well-defined control volume is assumed. In RTT mass of the system is described by setting ($b = 1$),

$$\frac{DM_s}{Dt} = \frac{\partial}{\partial t}\int_{cv} \rho dV + \int_{cs} \rho \mathbf{u} \cdot \hat{\mathbf{n}} dA. \tag{3.2}$$

Recalling that the system mass shall remain unchanged, the time change of mass equates to zero:

$$0 = \frac{\partial}{\partial t}\int_{cv} \rho dV + \int_{cs} \rho \mathbf{u} \cdot \hat{\mathbf{n}} dA. \tag{3.3}$$

The density of the fluid is assumed to be constant throughout the control volume, i.e. density may change in time but not for variation in position inside the control volume. The density as such is constant in regard to integration in spacial coordinates, and therefore not included in the volume integration,

$$0 = \frac{\partial}{\partial t}\left(\rho \int_{cv} 1 dV\right) + \rho \int_{cs} \mathbf{u} \cdot \hat{\mathbf{n}} dA. \tag{3.4}$$

The latter term regarding the flow/flux through the control surface includes the velocity vector and an area normal vector. Assuming the flow and the area normal vector to be parallel at all points where flow is crossing the surface leads to rewriting the integration as a summation of flow:

$$\rho \int_{cs} \mathbf{u} \cdot \hat{\mathbf{n}} dA = \rho \Sigma_{n=1}^{\mathrm{m}} u_n A_n. \tag{3.5}$$

Note that the area normal vector is outgoing on the control volume surface, which implies an outgoing flow to be positive and an incoming flow to be negative. The velocity u_n is the average velocity across the flow area A_n. We can hence combine the flows through the m areas in two parts, an outgoing and an incoming flow:

$$\rho \Sigma_{n=1}^{\mathrm{m}} u_n A_n = \rho \left(Q_{\mathrm{out}} - Q_{\mathrm{in}}\right) = \rho Q_{\mathrm{net}}, \tag{3.6}$$

where the Q_{out} and Q_{in} are the sum of outgoing and incoming flows respectively and Q_{net} is the total net out flow.

The first term in (3.4) regarding the control volume yields:

$$\frac{\partial}{\partial t}\left(\rho \int_{cv} 1 dV\right) = \frac{\partial}{\partial t}(\rho V). \tag{3.7}$$

I.e. the time derivative of the fluid density multiplied by the volume of the control volume. Applying the chain rule one gets,

$$\frac{\partial}{\partial t}(\rho V) = \rho\frac{\partial V}{\partial t} + \frac{\partial \rho}{\partial t}V = \rho\dot{V} + \dot{\rho}V. \tag{3.8}$$

Combining (3.6) and (3.8):

$$0 = \rho\left(Q_{\text{out}} - Q_{\text{in}}\right) + \rho\tfrac{\partial V}{\partial t} + \tfrac{\partial \rho}{\partial t}V, \tag{3.9}$$

$$\rho\left(Q_{\text{in}} - Q_{\text{out}}\right) = \rho\left(\tfrac{\partial V}{\partial t} + \tfrac{1}{\partial t}\tfrac{\partial \rho}{\rho}V\right), \tag{3.10}$$

and using (2.15) we see that;

$$\frac{\partial \rho}{\rho} = \frac{\partial p}{\beta}, \tag{3.11}$$

where (3.10) may be written as:

$$Q_{\text{in}} - Q_{\text{out}} = \frac{dV}{dt} + \frac{V}{\beta}\frac{dp}{dt}. \tag{3.12}$$

Equation (3.12) is named the flow continuity equation, however it is often denoted by the continuity equation. Note that we assumed the density to be constant within the control volume in spacial coordinates, as well as the pressure in the control volume, both may however vary in time. (3.12) is the flow continuity equation we use in lumped parameter modelling.

Example of Continuity Equation

Let's use the two "*tube systems*" in Fig. 3.1 as illustrations of how the flow continuity equation may be used for fluid power systems. First, the control volume is defined. The control volume limits are shown with the dashed lines in both (a) and (b). In (a) the control volume is fixed, contrary to (b) where the control volume will change if the piston on the right hand side moves.

Deriving a lumped parameter model, the pressure and density in the control volume is constant in spacial coordinates i.e. the pressure and density do not change with position in the control volume, only with time.

System (a) substituted into (3.12) yields:

$$Q_1 + Q_2 + Q_3 - Q_4 = 0 + \frac{V}{\beta}\dot{p}, \tag{3.13}$$

where Q_1, Q_2 and Q_3 are positive as they flow in and Q_4 is negative as it flows out. Note, that the volume is constant when the first right hand term is zero.

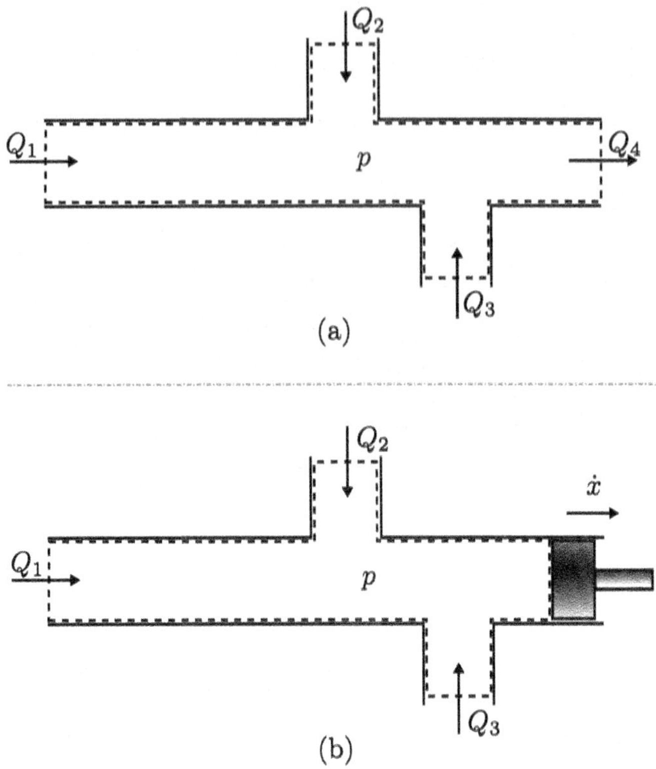

Fig. 3.1 Illustration of continuity equation with fixed and expanding control volumes

If the substitution into (3.12) is made for (b), one yields:

$$Q_1 + Q_2 + Q_3 = \dot{x}A + \frac{V_0 + xA}{\beta}\dot{p}, \tag{3.14}$$

where A is the area of the piston. Note that in system (b), the volume expansion is due to the movement of the piston and hence the control volume changes.

3.1.2 Continuity Equation—Differential Form

Models on differential form are models where the control volume is infinitesimally small. In our study we shall investigate a box fluid element placed in a flow field. The fluid element has the dimensions dx_1, dx_2, dx_3 and the fluid velocity in the flow field is:

$$\mathbf{u}(x_1, x_2, x_3, t) = \begin{bmatrix} u_1(x_1, x_2, x_3, t) \\ u_2(x_1, x_2, x_3, t) \\ u_3(x_1, x_2, x_3, t) \end{bmatrix}. \tag{3.15}$$

Notation-wise we will omit the parenthesis stating that the velocity is dependent on position and time; (x_1, x_2, x_3, t). Instead we shall use the notation: $\mathbf{u} = [u_1, u_2, u_3]$.

The rate of the mass change within the control volume on differential form may be written as,

$$\frac{\partial}{\partial t} \int_{cv} \rho dV \approx \frac{\partial \rho}{\partial t} dx_1 dx_2 dx_3. \tag{3.16}$$

For the mass flow across the control surface, the flow is separated in the three main directions x_1, x_2, x_3. The mass flow across the element boundary in direction x_1 is given as:

$$\begin{aligned} \text{Rate of outflow } x_1 = {} & (u_1\rho)(x_1 + dx_1, x_2, x_3)dx_2dx_3 \\ & -(u_1\rho)(x_1, x_2, x_3)dx_2dx_3. \end{aligned} \tag{3.17}$$

One may note that $(u_1\rho)(x_1 + dx_1, x_2, x_3)$ is the velocity in direction 1 and the density at the point $(x_1 + dx_1, x_2, x_3)$ and that dx_2dx_3 is the flow area perpendicular to direction 1 (the flow cross sectional area). By Taylor expansion the velocity in direction x_1 and the fluid density at point $(x_1 + dx_1,\ x_2,\ x_3)$ is given as:

$$(u_1\rho)(x_1 + dx_1, x_2, x_3) = (u_1\rho)(x_1, x_2, x_3) + \frac{\partial(u_1\rho)(x_1, x_2, x_3)}{\partial x_1} dx_1, \tag{3.18}$$

$$\frac{\partial(u_1\rho)(x_1, x_2, x_3)}{\partial x_1} dx_1 = (u_1\rho)(x_1 + dx_1, x_2, x_3) - (u_1\rho)(x_1, x_2, x_3). \tag{3.19}$$

With this, and likewise deviations for directions x_2, x_3, we yield:

$$\text{Out flow rate } x_1 = \frac{\partial(u_1\rho)}{\partial x_1} dx_1 dx_2 dx_3, \tag{3.20a}$$

$$\text{Out flow rate } x_2 = \frac{\partial(u_2\rho)}{\partial x_2} dx_2 dx_1 dx_3, \tag{3.20b}$$

$$\text{Out flow rate } x_3 = \frac{\partial(u_3\rho)}{\partial x_3} dx_3 dx_2 dx_1. \tag{3.20c}$$

The total net flow rate out of the fluid element is given as,

$$\text{Rate of net outflow} = \left(\frac{\partial(u_1\rho)}{\partial x_1} + \frac{\partial(u_2\rho)}{\partial x_2} + \frac{\partial(u_3\rho)}{\partial x_3} \right) dx_1 dx_2 dx_3. \tag{3.21}$$

Combining the rate of mass change within the fluid element (3.16) and the rate of net mass flow out (3.21) yields the continuity equation on differential form:

$$0 = \left(\frac{\partial \rho}{\partial t} + \frac{\partial (u_1 \rho)}{\partial x_1} + \frac{\partial (u_2 \rho)}{\partial x_2} + \frac{\partial (u_3 \rho)}{\partial x_3}\right) dx_1 dx_2 dx_3. \tag{3.22}$$

Removing the volume of the element from the equation and choosing conditions for unsteady compressible flow, steady compressible flow, and incompressible flow, will yield:

$$0 = \frac{\partial \rho}{\partial t} + \frac{\partial (u_1 \rho)}{\partial x_1} + \frac{\partial (u_2 \rho)}{\partial x_2} + \frac{\partial (u_3 \rho)}{\partial x_3}, \tag{3.23a}$$

$$0 = \frac{\partial (u_1 \rho)}{\partial x_1} + \frac{\partial (u_2 \rho)}{\partial x_2} + \frac{\partial (u_3 \rho)}{\partial x_3}, \tag{3.23b}$$

$$0 = \frac{\partial u_1}{\partial x_1} + \frac{\partial u_2}{\partial x_2} + \frac{\partial u_3}{\partial x_3}, \tag{3.23c}$$

respectively.

Cylindrical Coordinates

One may also have the continuity on differential form described in cylindrical coordinates. In cylindrical coordinates the continuity equations for unsteady compressible flow, steady compressible flow and incompressible flow, are given as:

$$0 = \frac{\partial \rho}{\partial t} + \frac{1}{r}\frac{\partial (r\rho u_r)}{\partial r} + \frac{1}{r}\frac{\partial (\rho u_\theta)}{\partial \theta} + \frac{\partial (\rho u_z)}{\partial z}, \tag{3.24a}$$

$$0 = \frac{1}{r}\frac{\partial (r\rho u_r)}{\partial r} + \frac{1}{r}\frac{\partial (\rho u_\theta)}{\partial \theta} + \frac{\partial (\rho u_z)}{\partial z}, \tag{3.24b}$$

$$0 = \rho\frac{1}{r}\frac{\partial (r u_r)}{\partial r} + \rho\frac{1}{r}\frac{\partial u_\theta}{\partial \theta} + \rho\frac{\partial u_z}{\partial z}, \tag{3.24c}$$

respectively.

3.2 Momentum of Fluids–Newton II. Law

This section looks at the momentum of fluids. We shall investigate the momentum change as a fluid flow through a system and how this momentum change may affect the components in which the momentum change occurs.

Newton's Second Law states that the change of momentum is equal to a resulting force. This may be written as:

$$\frac{D}{Dt}(M \cdot \mathbf{u}) = \Sigma \mathbf{F} \tag{3.25}$$

In the following subsections we shall derive the various terms, hence the momentum change and the resulting force, when dealing with fluids.

3.2.1 Differential Form—Cartesian Coordinates

Let's investigate a cube as fluid element. As we use Cartesian coordinates the side dimensions are chosen as dx_1, dx_2 and dx_3 as illustrated in Fig. 3.2.

In formulating Newton's Second Law for the fluid element in Fig. 3.3 the sum of forces acting on the element must be identified. The summation of forces is carried out direction-wise for directions x_1, x_2 and x_3. The forces acting on the element are categorised as either surface or body forces. The name indicates the point of attack of the forces, such as surface forces e.g. acts on the surface of the element.

Surface Forces

The surface forces are given as the stresses working on the surface fluid element multiplied by the area of the current surface. The normal forces on the element are found by summing the forces on the two opposing surfaces. For the first direction the resulting normal force is therefore given as,

$$F_{x_1,\sigma_{11}} = \sigma_{11}(x_1 + dx_1, x_2, x_3)dx_2dx_3 - \sigma_{11}(x_1, x_2, x_3)dx_2dx_3, \tag{3.26}$$

Where $\sigma_{11}\dot{()}$ is the normal stress in the fluid in direction 1 at the location given in the parenthesis. Investigating a cube, the surface areas perpendicular to direction 1 are equal at position x_1 and $x_1 + dx_1$, the normal force in direction 1 is the difference in normal stress at point $(x_1 + dx_1, x_2, x_3)$ and point (x_1, x_2, x_3) multiplied by the surface area dx_2dx_3:

$$F_{x_1,\sigma_{11}} = [\sigma_{11}(x_1 + dx_1, x_2, x_3) - \sigma_{11}(x_1, x_2, x_3)]\, dx_2dx_3. \tag{3.27}$$

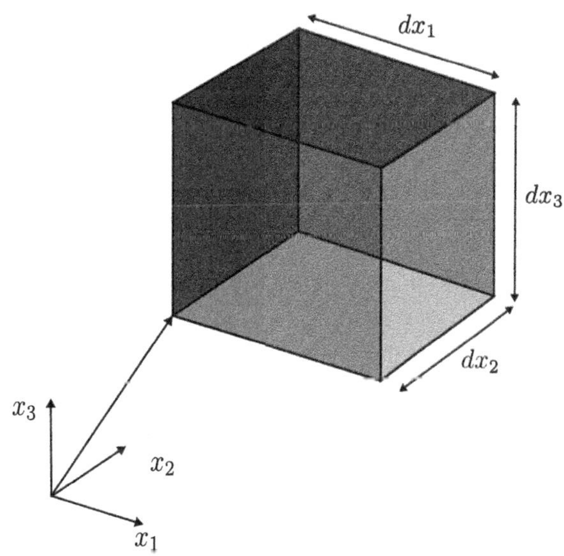

Fig. 3.2 Illustration of the investigated fluid element

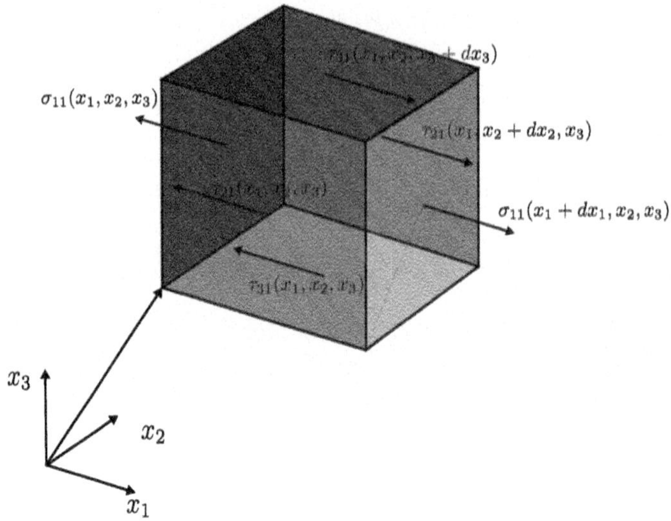

Fig. 3.3 Illustration of surface stresses on a fluid element. Only stresses in direction x_1 are shown

The stresses at the surface at point $(x_1 + dx_1, x_2, x_3)$ are found by Taylor expansion, where we neglect the higher order terms such that,

$$\sigma_{11}(x_1 + dx_1, x_2, x_3) = \sigma_{11}(x_1, x_2, x_3) + \frac{\partial \sigma_{11}(x_1, x_2, x_3)}{\partial x_1} dx_1 + \mathcal{O}(dx_1^2), \quad (3.28)$$

and as the force is calculated from the difference in normal stresses,

$$\sigma_{11}(x_1 + dx_1, x_2, x_3) - \sigma_{11}(x_1, x_2, x_3) = \frac{\partial \sigma_{11}(x_1, x_2, x_3)}{\partial x_1} dx_1, \quad (3.29)$$

is of interest such that the normal force in the first direction may be formulated as,

$$F_{x_1,\sigma_{11}} = \frac{\partial \sigma_{11}(x_1, x_2, x_3)}{\partial x_1} dx_1 dx_2 dx_3. \quad (3.30)$$

The resulting shear force in the first direction working on the x_1x_3 and x_1x_2 surfaces respectively are likewise found as:

$$F_{x_1,\tau_{21}} = \tau_{21}(x_1, x_2 + dx_2, x_3) dx_1 dx_3 - \tau_{21}(x_1, x_2, x_3) dx_1 dx_3, \quad (3.31)$$

and,

$$F_{x_1,\tau_{31}} = \tau_{31}(x_1, x_2, x_3 + dx_3) dx_1 dx_2 - \tau_{31}(x_1, x_2, x_3) dx_1 dx_2. \quad (3.32)$$

Utilising Taylor expansion once again to find the difference in stresses we yield:

$$\tau_{21}(x_1, x_2 + dx_2, x_3) = \tau_{21}(x_1, x_2, x_3) + \frac{\partial \tau_{21}(x_1, x_2, x_3)}{\partial x_2} dx_2 + \mathcal{O}(dx_2^2), \quad (3.33)$$

and neglecting higher order terms resulting in:

$$\tau_{21}(x_1, x_2 + dx_2, x_3) - \tau_{21}(x_1, x_2, x_3) = \frac{\partial \tau_{21}(x_1, x_2, x_3)}{\partial x_2} dx_2. \quad (3.34)$$

Likewise for τ_{31}:

$$\tau_{31}(x_1, x_2, x_3 + dx_3) = \tau_{31}(x_1, x_2, x_3) + \frac{\partial \tau_{31}(x_1, x_2, x_3)}{\partial x_3} dx_3 + \mathcal{O}(dx_3^2), \quad (3.35)$$

$$\tau_{31}(x_1, x_2, x_3 + dx_3) - \tau_{31}(x_1, x_2, x_3) = \frac{\partial \tau_{31}(x_1, x_2, x_3)}{\partial x_3} dx_3. \quad (3.36)$$

The shear forces working in the direction x_1 on the x_1x_3 and x_1x_2 surfaces respectively may then be found as:

$$F_{x_1,\tau_{21}} = \frac{\partial \tau_{21}(x_1, x_2, x_3)}{\partial x_2} dx_2 dx_1 dx_3, \quad (3.37)$$

and,

$$F_{x_1,\tau_{31}} = \frac{\partial \tau_{31}(x_1, x_2, x_3)}{\partial x_3} dx_3 dx_1 dx_2. \quad (3.38)$$

The total force due to surface stresses in direction x_1 summarises to:

$$F_{x_1} = F_{x_1,\sigma_{11}} + F_{x_1,\tau_{21}} + F_{x_1,\tau_{31}}, \quad (3.39)$$

$$F_{x_1} = \frac{\partial \sigma_{11}}{\partial x_1} dx_1 dx_2 dx_3 + \frac{\partial \tau_{21}}{\partial x_2} dx_2 dx_1 dx_3 + \frac{\partial \tau_{31}}{\partial x_3} dx_3 dx_1 dx_2, \quad (3.40)$$

organising with the volume of the element dV outside the parenthesis yield:

$$F_{x_1} = \left(\frac{\partial \sigma_{11}}{\partial x_1} + \frac{\partial \tau_{21}}{\partial x_2} + \frac{\partial \tau_{31}}{\partial x_3} \right) dx_1 dx_2 dx_3. \quad (3.41)$$

The same technique may be used to find forces in direction x_2 and x_3 due to surface stresses. Direction x_2:

$$F_{x_2} = F_{x_2,\sigma_{22}} + F_{x_2,\tau_{12}} + F_{x_2,\tau_{32}}, \quad (3.42)$$

$$F_{x_2} = \frac{\partial \sigma_{22}}{\partial x_2} dx_1 dx_2 dx_3 + \frac{\partial \tau_{12}}{\partial x_1} dx_2 dx_1 dx_3 + \frac{\partial \tau_{32}}{\partial x_3} dx_3 dx_1 dx_2, \quad (3.43)$$

$$F_{x_2} = \left(\frac{\partial \sigma_{22}}{\partial x_2} + \frac{\partial \tau_{12}}{\partial x_1} + \frac{\partial \tau_{32}}{\partial x_3} \right) dx_1 dx_2 dx_3. \quad (3.44)$$

and direction x_3:

$$F_{x_3} = F_{x_3,\sigma_{33}} + F_{x_3,\tau_{13}} + F_{x_3,\tau_{23}}, \tag{3.45}$$

$$F_{x_3} = \frac{\partial \sigma_{33}}{\partial x_3} dx_1 dx_2 dx_3 + \frac{\partial \tau_{13}}{\partial x_1} dx_2 dx_1 dx_3 + \frac{\partial \tau_{23}}{\partial x_2} dx_3 dx_1 dx_2, \tag{3.46}$$

$$F_{x_3} = \left(\frac{\partial \sigma_{33}}{\partial x_3} + \frac{\partial \tau_{13}}{\partial x_1} + \frac{\partial \tau_{23}}{\partial x_2} \right) dx_1 dx_2 dx_3. \tag{3.47}$$

In summary, the surface forces in the three directions are:

$$F_{x_1} = \left(\frac{\partial \sigma_{11}}{\partial x_1} + \frac{\partial \tau_{21}}{\partial x_2} + \frac{\partial \tau_{31}}{\partial x_3} \right) dx_1 dx_2 dx_3, \tag{3.48}$$

$$F_{x_2} = \left(\frac{\partial \sigma_{22}}{\partial x_2} + \frac{\partial \tau_{12}}{\partial x_1} + \frac{\partial \tau_{32}}{\partial x_3} \right) dx_1 dx_2 dx_3, \tag{3.49}$$

$$F_{x_3} = \left(\frac{\partial \sigma_{33}}{\partial x_3} + \frac{\partial \tau_{13}}{\partial x_1} + \frac{\partial \tau_{23}}{\partial x_2} \right) dx_1 dx_2 dx_3. \tag{3.50}$$

Note that the volume of the differential element is $dx_1 dx_2 dx_3 = dV$.

Surface forces on the element are seen to depend on how stress changes when the position is changed in the fluid flow field at hand. We shall later investigate how these stresses relate to parameters such as velocity and pressure in the flow field.

Body Forces

With the surface forces given above, the remaining forces to define are the body forces working on the element. Let's again separate the body forces into the three working directions $x_1,\ x_2,\ x_3$:

$$F_{b,x_1} = dmg_{x_1} + f_{\mathrm{b}}, x_1, \tag{3.51}$$

$$F_{b,x_2} = dmg_{x_2} + f_{\mathrm{b}}, x_2, \tag{3.52}$$

$$F_{b,x_3} = dmg_{x_3} + f_{\mathrm{b}}, x_3. \tag{3.53}$$

The first term for the body forces is the all-pervading gravitational force, given as the element mass multiplied by the gravitational acceleration in the three directions. The second term covers all other body forces. This could be e.g. forces on the element induced by a magnetic field. The gravitational force is, however, often the only body force of interest and even this may often be neglected in fluid power systems.

Acceleration

With the body and surface forces defined above, the last part of Newton's Second Law is the change of momentum in the element. The momentum change is non-zero if an acceleration or a mass change is present i.e. assuming the density to be constant, the momentum change is solely due to an acceleration which is given as the material derivative of the velocity field. With the velocity field given as $\mathbf{u} = [u_1,\ u_2,\ u_3]^{\mathrm{T}}$ the acceleration in direction 1 is:

$$a_1 = \frac{Du_1}{dt}, \tag{3.54}$$

which expands to:

$$\frac{Du_1}{dt} = \frac{\partial u_1}{\partial t} + \frac{\partial u_1}{\partial x_1}\frac{dx_1}{dt} + \frac{\partial u_1}{\partial x_2}\frac{dx_2}{dt} + \frac{\partial u_1}{\partial x_3}\frac{dx_3}{dt} \tag{3.55}$$

and may be written in terms of velocities in the three directions:

$$\frac{Du_1}{dt} = \frac{\partial u_1}{dt} + \frac{\partial u_1}{\partial x_1}u_1 + \frac{\partial u_1}{\partial x_2}u_2 + \frac{\partial u_1}{\partial x_3}u_3. \tag{3.56}$$

The acceleration vector including all three directions is therefore given as:

$$\mathbf{a} = \begin{bmatrix} a_1 \\ a_2 \\ a_3 \end{bmatrix} = \begin{bmatrix} \frac{\partial u_1}{\partial t} + u_1\frac{\partial u_1}{\partial x_1} + u_2\frac{\partial u_1}{\partial x_2} + u_3\frac{\partial u_1}{\partial x_3} \\ \frac{\partial u_2}{\partial t} + u_1\frac{\partial u_2}{\partial x_1} + u_2\frac{\partial u_2}{\partial x_2} + u_3\frac{\partial u_2}{\partial x_3} \\ \frac{\partial u_3}{\partial t} + u_1\frac{\partial u_3}{\partial x_1} + u_2\frac{\partial u_3}{\partial x_2} + u_3\frac{\partial u_3}{\partial x_3} \end{bmatrix}. \tag{3.57}$$

3.2.2 *Momentum Equation of a Fluid*

With the forces acting on the fluid element and the acceleration of the element described, Newton's second law may be formulated for the fluid element. Assuming a constant density also in time leaves $\frac{DM}{Dt} = 0$ such:

$$\frac{D\mathbf{u}}{Dt}M = \Sigma\mathbf{F}. \tag{3.58}$$

Substituting into (3.58) one yield:

$$\begin{bmatrix} dm\left(\frac{\partial u_1}{\partial t} + \frac{\partial u_1}{\partial x_1}u_1 + \frac{\partial u_1}{\partial x_2}u_2 + \frac{\partial u_1}{\partial x_3}u_3\right) \\ dm\left(\frac{\partial u_2}{\partial t} + \frac{\partial u_2}{\partial x_1}u_1 + \frac{\partial u_2}{\partial x_2}u_2 + \frac{\partial u_2}{\partial x_3}u_3\right) \\ dm\left(\frac{\partial u_3}{\partial t} + \frac{\partial u_3}{\partial x_1}u_1 + \frac{\partial u_3}{\partial x_2}u_2 + \frac{\partial u_3}{\partial x_3}u_3\right) \end{bmatrix} = \begin{bmatrix} dmg_{x_1} + \left(\frac{\partial \sigma_{11}}{\partial x_1} + \frac{\partial \tau_{21}}{\partial x_2} + \frac{\partial \tau_{31}}{\partial x_3}\right)dV \\ dmg_{x_2} + \left(\frac{\partial \sigma_{22}}{\partial x_2} + \frac{\partial \tau_{12}}{\partial x_1} + \frac{\partial \tau_{32}}{\partial x_3}\right)dV \\ dmg_{x_3} + \left(\frac{\partial \sigma_{33}}{\partial x_3} + \frac{\partial \tau_{13}}{\partial x_1} + \frac{\partial \tau_{23}}{\partial x_2}\right)dV \end{bmatrix}. \tag{3.59}$$

Utilising that $\rho = \frac{dm}{dV}$ and that the volume of the element is non-zero, the general partial differential equation of motion for a fluid is given as:

$$\rho g_{x_1} + \left(\frac{\partial \sigma_{11}}{\partial x_1} + \frac{\partial \tau_{21}}{\partial x_2} + \frac{\partial \tau_{31}}{\partial x_3}\right) = \rho\left(\frac{\partial u_1}{\partial t} + \frac{\partial u_1}{\partial x_1}u_1 + \frac{\partial u_1}{\partial x_2}u_2 + \frac{\partial u_1}{\partial x_3}u_3\right), \tag{3.60a}$$

$$\rho g_{x_2} + \left(\frac{\partial \sigma_{22}}{\partial x_2} + \frac{\partial \tau_{12}}{\partial x_1} + \frac{\partial \tau_{32}}{\partial x_3}\right) = \rho\left(\frac{\partial u_2}{\partial t} + \frac{\partial u_2}{\partial x_1}u_1 + \frac{\partial u_2}{\partial x_2}u_2 + \frac{\partial u_2}{\partial x_3}u_3\right), \tag{3.60b}$$

$$\rho g_{x_3} + \left(\frac{\partial \sigma_{33}}{\partial x_3} + \frac{\partial \tau_{13}}{\partial x_1} + \frac{\partial \tau_{23}}{\partial x_2}\right) = \rho\left(\frac{\partial u_3}{\partial t} + \frac{\partial u_3}{\partial x_1}u_1 + \frac{\partial u_3}{\partial x_2}u_2 + \frac{\partial u_3}{\partial x_3}u_3\right), \tag{3.60c}$$

Note that forces are gathered on the left side and momentum changes are given on the right hand side. In vector notation Newton's second law for a fluid element may be written as:

$$\rho\mathbf{g} + \nabla \cdot \mathbf{T} = \rho\left(\frac{d\mathbf{u}}{dt} + (\mathbf{u}\cdot\nabla)\mathbf{u}\right), \tag{3.61}$$

where the stress tensor, **T**, is given as:

$$\mathbf{T} = \begin{bmatrix} \sigma_{11} & \tau_{12} & \tau_{13} \\ \tau_{21} & \sigma_{22} & \tau_{23} \\ \tau_{31} & \tau_{32} & \sigma_{33}, \end{bmatrix} \tag{3.62}$$

and $(\mathbf{u}\cdot\nabla)\mathbf{u}$ expands to:

$$(\mathbf{u}\cdot\nabla)\mathbf{u} = \left([u_1\ u_2\ u_3]\cdot\begin{bmatrix}\frac{\partial}{\partial x_1}\\ \frac{\partial}{\partial x_2}\\ \frac{\partial}{\partial x_3}\end{bmatrix}\right)\mathbf{u} = \left(u_1\frac{\partial}{\partial x_1} + u_2\frac{\partial}{\partial x_2} + u_2\frac{\partial}{\partial x_3}\right)\begin{bmatrix}u_1\\ u_2\\ u_3\end{bmatrix}, \tag{3.63}$$

and may be compiled to:

$$(\mathbf{u}\cdot\nabla)\mathbf{u} = \begin{bmatrix} u_1\frac{\partial u_1}{\partial x1} + u_2\frac{\partial u_1}{\partial x2} + u_3\frac{\partial u_1}{\partial x3} \\ u_1\frac{\partial u_2}{\partial x1} + u_2\frac{\partial u_2}{\partial x2} + u_3\frac{\partial u_2}{\partial x3} \\ u_1\frac{\partial u_3}{\partial x1} + u_2\frac{\partial u_3}{\partial x2} + u_3\frac{\partial u_3}{\partial x3} \end{bmatrix}. \tag{3.64}$$

Note that the density is assumed constant, hence we are dealing with an incompressible fluid. The reader is encouraged to study the effect of varying density in other materials , however, this is seldom relevant for simple lumped parameter models of fluid power systems and components.

3.2.3 *Conservation of Momentum—Control Volume Form*

Using a control volume approach leads to employing Reynolds Transport Theorem,

$$\frac{D}{Dt}B_{\text{sys}} = \frac{\partial}{\partial t}\int_{\text{cv}} \rho b dV + \int_{\text{cs}} \rho b\bar{u}\cdot\hat{n}\, dA. \tag{3.65}$$

Investigating momentum, the parameter b must be the velocity since $B_{\text{sys}} = mb$. The change in momentum in the system is therefore given as:

$$\frac{D}{Dt}(mu) = \frac{\partial}{\partial t}\int_{\text{cv}} \rho\bar{u}\cdot dV + \int_{\text{cs}} \rho u\bar{u}\cdot\hat{n}\, dA, \tag{3.66}$$

where the "system" is the fluid in the control volume. One may note that the momentum change is given in two terms, one describing the momentum change in the fluid inside the control volume, and one describing the momentum change across the

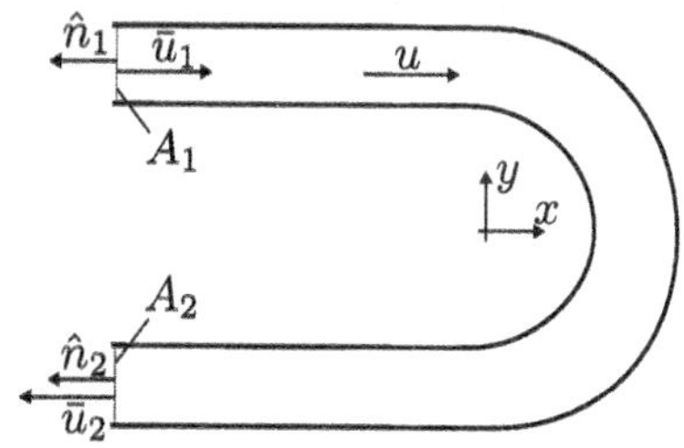

Fig. 3.4 Flow through a u-turn

boundary or surface of the control volume. The change in momentum is driven by the sum of external forces acting on the control volume as body forces and surface forces and may as such be written as,

$$\frac{\mathbf{DuM}}{Dt} = \Sigma\mathbf{F}. \tag{3.67}$$

Force due to flow u-turn

Let's examine the simple case of a steady flow through a u-turn, see Fig. 3.4. Let's assume the flow to be steady in time and inviscid, hence a constant flow rate and no shearing stresses from the walls and between the fluid layers exists. A flow Q is flowing through a pipe with inner diameter d.

With a control volume bounded by the inner wall of the tube and the two dotted horizontal lines, a momentum equation may be written as:

$$\Sigma\mathbf{F} = \frac{\partial}{\partial t}\int_{\mathrm{cv}} \rho\bar{u}dV + \int_{\mathrm{cs}} \rho u\bar{u}\cdot\hat{n}\, dA. \tag{3.68}$$

In the current case the flow is steady in time where the first term is zero, as the time derivative is zero. For the second term we see two surface areas where flow is crossing the surface. Both surface crossings are in the $x-$direction, hence;

$$\Sigma F_{\mathrm{x}} = \rho u_1 u_1(-A_1) + \rho(-u_2)(u_2)(A_2), \tag{3.69}$$

$$\Sigma F_{\mathrm{x}} = -\rho(u_1^2 A_1 + u_2^2 A_2) \tag{3.70}$$

As the areas are equal $A_1 = A_2$ the flow velocity is equal, and the force is:

$$\Sigma F_{\mathrm{x}} = -2\rho u_1^2 A_1 = -2\rho\frac{Q^2}{A_1} = -8\rho\frac{Q^2}{d^2\pi}. \tag{3.71}$$

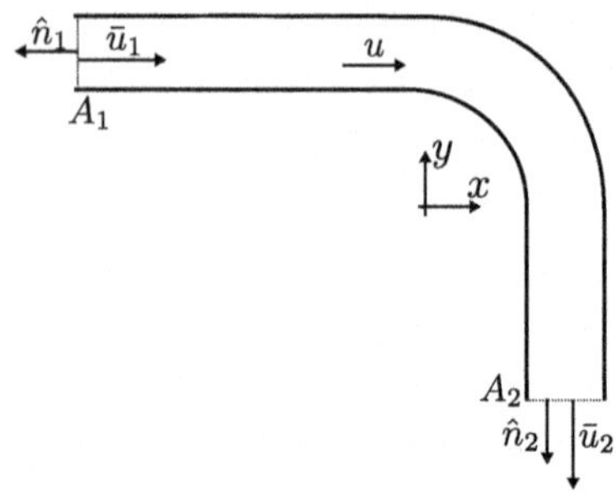

Fig. 3.5 Flow through a 90 ° turn

Based on this brief equating the force required to hold the u-bend in place is given. Note that the computed force is a force required to perform the spacial acceleration of the fluid. Here we see that if a fluid flow changes direction, the momentum of the fluid changes and a force is required to perform this momentum change even in case of a steady flow.

3.2.3.1 Force Due to Flow L-Turn

Let's examine another simple case of steady flow through a 90 ° turn, see Fig. 3.5. We again assume the flow to be steady in time and inviscid.

Once more we define the control volume bounds as the inner wall and the two dotted lines. In the current case however the in flow is in the x-direction and the out flow is in the y-direction. As such, the forces are summed in both the x and y direction respectively.

$$\Sigma F_x = \rho u_1 u_1(-A_1) = -4\rho \tfrac{Q^2}{d^2\pi}, \tag{3.72}$$

$$\Sigma F_y = \rho(-u_2)(-u_2)(-A_2) = -4\rho \tfrac{Q^2}{d^2\pi}. \tag{3.73}$$

To hold the 90 ° bend in place a resulting force $[-4\rho\frac{Q^2}{d^2\pi},\ -4\rho\frac{Q^2}{d^2\pi},\ 0]$ is required. Note that the force required to hold the L-turn is smaller than the force required to hold the u-turn, as the scalar of the force is:

$$|\mathbf{F}| = \sqrt{\left(-4\rho\frac{Q^2}{d^2\pi}\right)^2 + \left(-4\rho\frac{Q^2}{d^2\pi}\right)^2} = \sqrt{32}\rho\frac{Q^2}{d^2\pi} = 5.66\rho\frac{Q^2}{d^2\pi}. \tag{3.74}$$

3.3 Euler's Equations of Motion–Inviscid Flow

In the case of an inviscid fluid flow the stress tensor **T** in (3.61) is greatly simplified, since the shearing stresses are zero and the normal stresses are equal to the pressure, $-p = \sigma_{11} = \sigma_{22} = \sigma_{33}$. As such, a fluid element in compression has a positive pressure. Substituting this in (3.60) yields:

$$\rho g_{x_1} - \frac{\partial p}{\partial x_1} = \rho \left(\frac{\partial u_1}{\partial t} + \frac{\partial u_1}{\partial x_1} u_1 + \frac{\partial u_1}{\partial x_2} u_2 + \frac{\partial u_1}{\partial x_3} u_3 \right), \tag{3.75}$$

$$\rho g_{x_2} - \frac{\partial p}{\partial x_2} = \rho \left(\frac{\partial u_2}{\partial t} + \frac{\partial u_2}{\partial x_1} u_1 + \frac{\partial u_2}{\partial x_2} u_2 + \frac{\partial u_2}{\partial x3} u_3 \right), \tag{3.76}$$

$$\rho g_{x_3} - \frac{\partial p}{\partial x_3} = \rho \left(\frac{\partial u_3}{\partial t} + \frac{\partial u_3}{\partial x_1} u_1 + \frac{\partial u_3}{\partial x_2} u_2 + \frac{\partial u_3}{\partial x_3} u_3 \right). \tag{3.77}$$

These three equations are referred to as Euler's Equations of Motion. These nonlinear partial differential equations describe the pressure and velocity of an inviscid fluid.

3.4 Viscous Flow

For Newtonian fluids, the surface stresses are linearly dependent on the rate of strain with viscosity as a proportionality constant. The stress tensor for Newtonian fluids is given as:

$$T_{ij} = -\left(p_m + \tfrac{2}{3}\mu \nabla \cdot \mathbf{u}\right) \delta_{ij} + \mu \left(\frac{\partial u_i}{\partial x_j} + \frac{\partial u_j}{\partial x_i} \right), \tag{3.78}$$

where

$$p_m = p - \lambda \nabla \cdot \mathbf{u}, \tag{3.79}$$

and,

$$\delta_{ij} = 1 \;\text{ for } i = j \quad \text{and} \quad \delta_{ij} = 0 \;\text{ for } i \neq j. \tag{3.80}$$

In Cartesian coordinates the stress tensor is given as:

$$\mathbf{T} = \begin{bmatrix} \sigma_{11} & \tau_{12} & \tau_{13} \\ \tau_{21} & \sigma_{22} & \tau_{23} \\ \tau_{31} & \tau_{32} & \sigma_{33}. \end{bmatrix} \tag{3.81}$$

Writing the individual elements based on (3.78) yields:

$$\sigma_{11} = -p - \tfrac{2}{3}\mu\left(\tfrac{\partial u_1}{\partial x_1} + \tfrac{\partial u_2}{\partial x_2} + \tfrac{\partial u_3}{\partial x_3}\right) + 2\mu\tfrac{\partial u_1}{\partial x_1}, \tag{3.82a}$$

$$\sigma_{22} = -p - \tfrac{2}{3}\mu\left(\tfrac{\partial u_1}{\partial x_1} + \tfrac{\partial u_2}{\partial x_2} + \tfrac{\partial u_3}{\partial x_3}\right) + 2\mu\tfrac{\partial u_2}{\partial x_2}, \tag{3.82b}$$

$$\sigma_{33} = -p - \tfrac{2}{3}\mu\left(\tfrac{\partial u_1}{\partial x_1} + \tfrac{\partial u_2}{\partial x_2} + \tfrac{\partial u_3}{\partial x_3}\right) + 2\mu\tfrac{\partial u_3}{\partial x_3}, \tag{3.82c}$$

$$\tau_{12} = \tau_{21} = \mu\left(\tfrac{\partial u_1}{\partial x_2} + \tfrac{\partial u_2}{\partial x_1}\right), \tag{3.82d}$$

$$\tau_{23} = \tau_{32} = \mu\left(\tfrac{\partial u_2}{\partial x_3} + \tfrac{\partial u_3}{\partial x_2}\right), \tag{3.82e}$$

$$\tau_{31} = \tau_{13} = \mu\left(\tfrac{\partial u_3}{\partial x_1} + \tfrac{\partial u_1}{\partial x_3}\right). \tag{3.82f}$$

3.4.1 Navier-Stokes Equations—Incompressible Fluid

Considering flow in an incompressible Newtonian fluid, the volumetric strain equals zero and the stress tensor simplifies (by using (3.23c)) to:

$$\sigma_{11} = -p + 2\mu\tfrac{\partial u_1}{\partial x_1}, \tag{3.83a}$$

$$\sigma_{22} = -p + 2\mu\tfrac{\partial u_2}{\partial x_2}, \tag{3.83b}$$

$$\sigma_{33} = -p + 2\mu\tfrac{\partial u_3}{\partial x_3}, \tag{3.83c}$$

and,

$$\tau_{12} = \tau_{21} = \mu\left(\tfrac{\partial u_1}{\partial x_2} + \tfrac{\partial u_2}{\partial x_1}\right), \tag{3.83d}$$

$$\tau_{23} = \tau_{32} = \mu\left(\tfrac{\partial u_2}{\partial x_3} + \tfrac{\partial u_3}{\partial x_2}\right), \tag{3.83e}$$

$$\tau_{31} = \tau_{13} = \mu\left(\tfrac{\partial u_3}{\partial x_1} + \tfrac{\partial u_1}{\partial x_3}\right). \tag{3.83f}$$

The normal stresses depend only on the pressure and the rate of velocity in the current direction. On the other hand the shear stress depends on the rate of velocity in the orthogonal direction.

With the stress tensor known, the surface forces may be inserted in the momentum equation for a fluid. Let's examine the surface forces firstly in the normal direction and secondly in the shearing direction.

Normal Direction

The normal force is given as the normal stress multiplied by the area on which it is acting. However, as we saw in going from Eq. (3.59) to (3.60) we divided by the element volume dV leading to the term $\frac{\partial \sigma_{ii}}{\partial x_i}$. The change in normal stresses is in the three directions given as:

$$\frac{\partial \sigma_{11}}{\partial x_1} = -\frac{\partial}{\partial x_1} p + \frac{\partial}{\partial x_1}\left(2\mu \frac{\partial u_1}{\partial x_1}\right) = -\frac{\partial p}{\partial x_1} + \left(2\mu \frac{\partial^2 u_1}{\partial x_1^2}\right), \tag{3.84a}$$

$$\frac{\partial \sigma_{22}}{\partial x_2} = -\frac{\partial}{\partial x_2} p + \frac{\partial}{\partial x_2}\left(2\mu \frac{\partial u_2}{\partial x_2}\right) = -\frac{\partial p}{\partial x_2} + \left(2\mu \frac{\partial^2 u_2}{\partial x_2^2}\right), \tag{3.84b}$$

$$\frac{\partial \sigma_{33}}{\partial x_3} = -\frac{\partial}{\partial x_3} p + \frac{\partial}{\partial x_3}\left(2\mu \frac{\partial u_3}{\partial x_3}\right) = -\frac{\partial p}{\partial x_3} + \left(2\mu \frac{\partial^2 u_3}{\partial x_3^2}\right). \tag{3.84c}$$

Shearing Direction

The shear forces are given as the shear stress elements multiplied by the area on which they are acting. Once again the derivation step from Eq. (3.59) to (3.60) yields that the changes in shear stress required are:

$$\frac{\partial \tau_{12}}{\partial x_1} = \mu\left(\frac{\partial}{\partial x_1}\frac{\partial u_1}{\partial x_2} + \frac{\partial}{\partial x_1}\frac{\partial u_2}{\partial x_1}\right) = \mu\left(\frac{\partial^2 u_1}{\partial x_1 \partial x_2} + \frac{\partial^2 u_2}{\partial x_1^2}\right), \tag{3.85a}$$

$$\frac{\partial \tau_{21}}{\partial x_2} = \mu\left(\frac{\partial}{\partial x_2}\frac{\partial u_1}{\partial x_2} + \frac{\partial}{\partial x_2}\frac{\partial u_2}{\partial x_1}\right) = \mu\left(\frac{\partial^2 u_1}{\partial^2 x_2} + \frac{\partial^2 u_2}{\partial x_1 \partial x_2}\right), \tag{3.85b}$$

and,

$$\frac{\partial \tau_{23}}{\partial x_2} = \mu\left(\frac{\partial}{\partial x_2}\frac{\partial u_2}{\partial x_3} + \frac{\partial}{\partial x_2}\frac{\partial u_3}{\partial x_2}\right) = \mu\left(\frac{\partial^2 u_2}{\partial x_2 \partial x_3} + \frac{\partial^2 u_3}{\partial x_2^2}\right), \tag{3.85c}$$

$$\frac{\partial \tau_{32}}{\partial x_3} = \mu\left(\frac{\partial}{\partial x_3}\frac{\partial u_2}{\partial x_3} + \frac{\partial}{\partial x_3}\frac{\partial u_3}{\partial x_2}\right) = \mu\left(\frac{\partial^2 u_2}{\partial x_3^2} + \frac{\partial^2 u_3}{\partial x_2 \partial x_3}\right), \tag{3.85d}$$

and,

$$\frac{\partial \tau_{31}}{\partial x_3} = \mu\left(\frac{\partial}{\partial x_3}\frac{\partial u_3}{\partial x_1} + \frac{\partial}{\partial x_3}\frac{\partial u_1}{\partial x_3}\right) = \mu\left(\frac{\partial^2 u_3}{\partial x_1 \partial x_3} + \frac{\partial^2 u_1}{\partial x_3^2}\right), \tag{3.85e}$$

$$\frac{\partial \tau_{13}}{\partial x_1} = \mu\left(\frac{\partial}{\partial x_1}\frac{\partial u_3}{\partial x_1} + \frac{\partial}{\partial x_1}\frac{\partial u_1}{\partial x_3}\right) = \mu\left(\frac{\partial^2 u_3}{\partial x_1^2} + \frac{\partial^2 u_1}{\partial x_1 \partial x_3}\right). \tag{3.85f}$$

Substituting the elements of the stress tensor for an incompressible fluid into the general equation of motion of a fluid, (3.60) yields for direction 1:

$$\begin{aligned} &\rho\left(\frac{\partial u_1}{\partial t} + u_1\frac{\partial u_1}{\partial x_1} + u_2\frac{\partial u_1}{\partial x_2} + u_3\frac{\partial u_1}{\partial x_3}\right) = \rho g_{x_1} + \\ &\left(-\frac{\partial p}{\partial x_1} + \left(2\mu\frac{\partial^2 u_1}{\partial x_1^2}\right) + \mu\left(\frac{\partial^2 u_1}{\partial^2 x_2} + \frac{\partial^2 u_2}{\partial x_1 \partial x_2}\right) + \mu\left(\frac{\partial^2 u_3}{\partial x_1 \partial x_3} + \frac{\partial^2 u_1}{\partial x_3^2}\right)\right) \\ &= \rho g_{x_1} - \frac{\partial p}{\partial x_1} + \mu\left(2\frac{\partial^2 u_1}{\partial x_1^2} + \frac{\partial^2 u_1}{\partial^2 x_2} + \frac{\partial^2 u_1}{\partial x_3^2}\right) + \mu\left(\frac{\partial^2 u_2}{\partial x_1 \partial x_2} + \frac{\partial^2 u_3}{\partial x_1 \partial x_3}\right). \end{aligned} \tag{3.86}$$

Recall equation (3.23c) for incompressible fluid flow which yields:

$$\left(\frac{\partial^2 u_2}{\partial x_1 \partial x_2} + \frac{\partial^2 u_3}{\partial x_1 \partial x_3}\right) = -\frac{\partial^2 u_1}{\partial x_1^2}. \tag{3.87}$$

Expanding to all three directions yields the Navier-Stokes equations for an incompressible fluid flow:

$$\rho\left(\frac{\partial u_1}{\partial t} + u_1\frac{\partial u_1}{\partial x_1} + u_2\frac{\partial u_1}{\partial x_2} + u_3\frac{\partial u_1}{\partial x_3}\right) = \rho g_{x_1} - \frac{\partial p}{\partial x_1} + \mu\left(\frac{\partial^2 u_1}{\partial x_1^2} + \frac{\partial^2 u_1}{\partial x_2^2} + \frac{\partial^2 u_1}{\partial x_3^2}\right), \tag{3.88a}$$

$$\rho\left(\frac{\partial u_2}{\partial t} + u_1\frac{\partial u_2}{\partial x_1} + u_2\frac{\partial u_2}{\partial x_2} + u_3\frac{\partial u_2}{\partial x_3}\right) = \rho g_{x_2} - \frac{\partial p}{\partial x_2} + \mu\left(\frac{\partial^2 u_2}{\partial x_1^2} + \frac{\partial^2 u_2}{\partial x_2^2} + \frac{\partial^2 u_2}{\partial x_3^2}\right), \tag{3.88b}$$

$$\rho\left(\frac{\partial u_3}{\partial t} + u_1\frac{\partial u_3}{\partial x_1} + u_2\frac{\partial u_3}{\partial x_2} + u_3\frac{\partial u_3}{\partial x_3}\right) = \rho g_{x_3} - \frac{\partial p}{\partial x_3} + \mu\left(\frac{\partial^2 u_3}{\partial x_1^2} + \frac{\partial^2 u_3}{\partial x_2^2} + \frac{\partial^2 u_3}{\partial x_3^2}\right). \tag{3.88c}$$

Navier-Stokes equations are here derived from conservation of momentum. Furthermore, a constitutive law for Newtonian fluids are utilised and finally the fluid was assumed incompressible, yielding a constant density in both spacial coordinates and in time.

3.4.1.1 Cylindrical Coordinates

In cylindrical coordinates the Navier-Stokes equations read:

$$\begin{aligned}\rho\left(\frac{\partial u_r}{\partial t} + u_r\frac{\partial u_r}{\partial r} + \frac{u_\theta}{r}\frac{\partial u_r}{\partial \theta} - \frac{u_\theta^2}{r} + u_z\frac{\partial u_r}{\partial z}\right) &= \rho g_r - \frac{\partial p}{\partial r}\\ &+\mu\left(\frac{1}{r}\frac{\partial}{\partial r}\left(r\frac{\partial u_r}{\partial r}\right) - \frac{u_r}{r^2} + \frac{1}{r^2}\frac{\partial u_r^2}{\partial \theta^2} - \frac{2}{r^2}\frac{\partial u_\theta}{\partial \theta} + \frac{\partial^2 u_r}{\partial z^2}\right),\end{aligned} \tag{3.89a}$$

$$\begin{aligned}\rho\left(\frac{\partial u_\theta}{\partial t} + u_r\frac{\partial u_\theta}{\partial r} + \frac{u_\theta}{r}\frac{\partial u_\theta}{\partial \theta} + \frac{u_\theta u_r}{r} + u_z\frac{\partial u_\theta}{\partial x_z}\right) &= \rho g_\theta - \frac{1}{r}\frac{\partial p}{\partial \theta}\\ &+\mu\left(\frac{1}{r}\frac{\partial}{\partial r}\left(r\frac{\partial u_\theta}{\partial r}\right) - \frac{u_\theta}{r^2} + \frac{1}{r^2}\frac{\partial^2 u_\theta}{\partial \theta^2} + \frac{2}{r^2}\frac{\partial u_r}{\partial \theta} + \frac{\partial^2 u_\theta}{\partial z^2}\right),\end{aligned} \tag{3.89b}$$

$$\begin{aligned}\rho\left(\frac{\partial u_z}{\partial t} + u_r\frac{\partial u_z}{\partial r} + \frac{u_\theta}{r}\frac{\partial u_z}{\partial \theta} + u_z\frac{\partial u_z}{\partial z}\right) &= \rho g_z - \frac{\partial p}{\partial z}\\ &+\mu\left(\frac{1}{r}\frac{\partial}{\partial r}\left(r\frac{\partial u_z}{\partial r}\right) + \frac{1}{r^2}\frac{\partial^2 u_z}{\partial \theta^2} + \frac{\partial^2 u_z}{\partial z^2}\right).\end{aligned} \tag{3.89c}$$

Chapter 4
Flow Through Restriction

Abstract This chapter introduces models for steady fluid flow. Firstly, steady flow through smooth pipes and hoses are derived from a laminar flow dominated solely by shear forces. Secondly, the experimentally derived model for turbulent flow is introduced. A model useful for laminar leakage flow through small gaps is derived next, followed by a derivation of the orifice equation.

In fluid power systems, restrictions are often used to control the fluid flow. In the coming chapters we shall learn how valves restrict or guide fluid flow such that the design engineering gets the requested system performance. Various fittings, pipes and hoses are used to distribute fluid from component to component, however, these paths also induce some flow resistance and as such are discussed in this chapter.

In the current chapter we shall develop steady state models for fluid flow through various types of restrictions, e.g. pipes and hoses, gaps and orifices.

4.1 Turbulent or Laminar—Reynolds Number

Fluid flow may have various flow regimes characterised by the the movement of the individual fluid particle in the flow field. In laminar flow the streamlines for individual particles are neatly layered and flow "parallel". Contrary turbulent flow features a chaotic streamline pattern for individual particles. The two flow regimes are illustrated in Fig. 4.1.

The laminar streamline pattern appears as viscous forces dominate in the fluid flow while inertia forces dominate in turbulent flows.

Reynolds number is the ratio of inertia to viscous forces in a fluid flow. This may also be given as the ratio between momentum of the fluid and the shearing stress in the fluid flow such that:

$$Re = \frac{\text{inertial forces}}{\text{viscous forces}} = \frac{\rho \mathbf{u} L}{\mu}, \tag{4.1}$$

A. H. Hansen, *Fluid Power Systems*, Fluid Mechanics and Its Applications 129,
https://doi.org/10.1007/978-3-031-15089-0_4

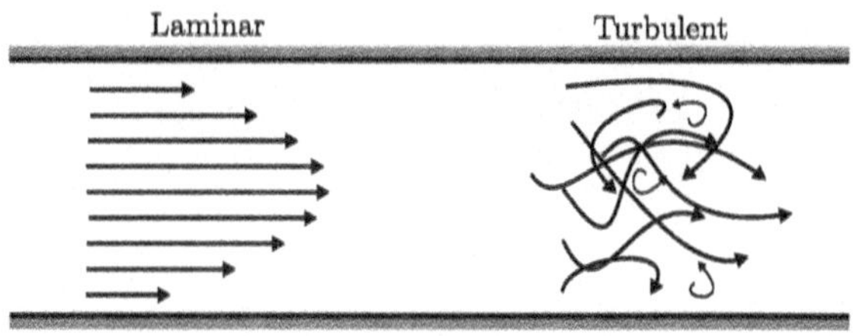

Fig. 4.1 Illustration of particle paths in a laminar and in a turbulent fluid flow

where ρ and μ are the density and dynamic viscosity of the fluid respectively, while $\mathbf{u}$ and L are the mean fluid velocity and characteristic length respectively. Reynolds number may also be set up as the ratio between dynamic pressure and shear stress such as:

$$Re = \frac{\text{dynamic pressure}}{\text{Shear stress}} = \frac{\rho \mathbf{u}^2}{\mu \frac{\mathbf{u}}{L}} = \frac{\rho \mathbf{u} L}{\mu}. \tag{4.2}$$

Reynolds number indicates which forces, inertia or shearing, are most dominant in the fluid flow. A low number indicates that viscous forces dominate, whereas a high Reynolds number indicates that inertia forces dominate.

4.2 Flow in a Pipe

In fluid power systems, pipes and hoses connecting components are normally circular in cross-section. We therefore assume a circular cross-section of pipes and hoses.

Developing a model for the fluid flow in a pipe the flow regime is, as previously introduced, determined with Reynolds number,

$$Re = \frac{\rho \mathbf{u} D_{\mathrm{H}}}{\mu}, \tag{4.3}$$

where ρ and μ are the fluid density and dynamic viscosity respectively. $\mathbf{u}$ is the mean fluid velocity and D_{H} is the hydraulic diameter. Note that for flows in closed pipes the hydraulic diameter is used as characteristic length. The hydraulic diameter is for shapes such as circles, squares, rectangles or annular ducts with height and width in the same order given as:

$$D_{\mathrm{h}} = \frac{4A}{P}, \tag{4.4}$$

where A is the cross-sectional flow area and P is the wetted perimeter. Hence, the perimeter of all channel walls touching the fluid in the cross-section.

For a circular flow channel with diameter d the hydraulic diameter is given as:

$$D_\mathrm{h} = \frac{4A}{P} = \frac{4d^2\frac{\pi}{4}}{d\pi} = d. \tag{4.5}$$

Reynolds number for a fluid flow in the circular tube is then:

$$Re = \frac{\rho \mathbf{u} d}{\mu} = \frac{\rho Q d}{A\mu} = \frac{\rho Q d}{d^2\frac{\pi}{4}\mu} = \frac{4\rho Q}{d\pi\mu}, \tag{4.6}$$

or

$$Re = \frac{\mathbf{u} d}{\nu} = \frac{Qd}{A\nu} = \frac{Qd}{d^2\frac{\pi}{4}\nu} = \frac{4Q}{d\pi\nu}. \tag{4.7}$$

Through empirical tests Reynolds discovered that for "pipe flow":

1. If $Re < 2000$, flow is laminar.
2. If $Re > 2400$, flow is turbulent.

Between the two values the flow may be in a mixed transition state. However, during analysis we shall use 2300 as the limit for both, and as such we say the flow is either purely laminar or turbulent when under and above 2300 respectively.

4.2.1 From Navier-Stokes Equation

Let's use Navier-Stokes equation as offset in deriving a formula for the fluid flow through a pipe. We use cylindrical coordinates and place the coordinate system as in Fig. 4.2 with the origin at the centre of the pipe cross-section and with z in the flow direction.

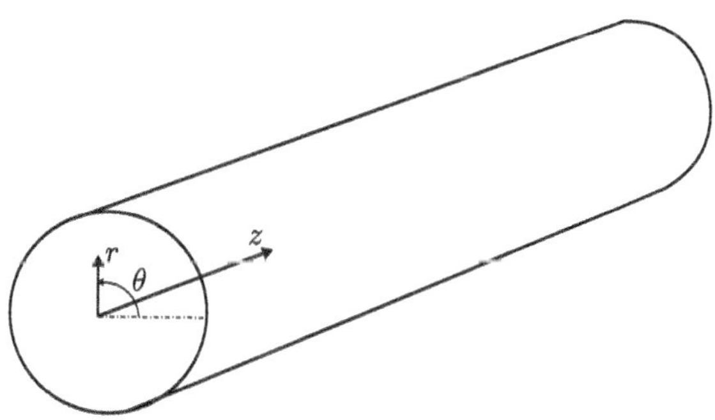

Fig. 4.2 Illustration of tube, with cylindrical coordinates system

The flow of interest is in the z-direction where we employ (3.89c) reprinted below for ease of reading.

$$\rho\left(\frac{\partial u_z}{\partial t}+u_r\frac{\partial u_z}{\partial r}+\frac{u_\theta}{r}\frac{\partial u_z}{\partial \theta}+u_z\frac{\partial u_z}{\partial z}\right)=\rho g_z-\frac{\partial p}{\partial z}+\mu\left(\frac{1}{r}\frac{\partial}{\partial r}\left(r\frac{\partial u_z}{\partial r}\right)+\frac{1}{r^2}\frac{\partial^2 u_z}{\partial \theta^2}+\frac{\partial^2 u_z}{\partial z^2}\right). \tag{4.8}$$

We assume that there is no fluid velocity in the r and θ direction, and we yield:

$$u_r = u_\theta = 0. \tag{4.9}$$

Hence, there is no flow towards the wall and no flow circling the z-axis. The flow through the pipe is further assumed to be symmetrical around the z-axis, this imposes:

$$\frac{\partial u_z}{\partial \theta}=0, \qquad \frac{\partial^2 u_z}{\partial \theta^2}=0. \tag{4.10}$$

The last spacial assumptions are that the flow does not accelerate along the length of the tube and the gravitational term is neglected.

$$\frac{\partial u_z}{\partial z}=0, \qquad \frac{\partial^2 u_z}{\partial z^2}=0, \qquad \rho g_z=0. \tag{4.11}$$

With these assumptions we yield a much simplified version of Navier-Stokes equation,

$$\rho\left(\frac{\partial u_z}{\partial t}\right)=-\frac{\partial p}{\partial z}+\mu\left(\frac{1}{r}\frac{\partial}{\partial r}\left(r\frac{\partial u_z}{\partial r}\right)\right). \tag{4.12}$$

If one assumes the flow to be steady, the time derivative will be zero, such that a constant flow is led through the pipe.

A steady flow with velocity components only in the z-direction is given as:

$$0=-\frac{\partial p}{\partial z}+\mu\left(\frac{1}{r}\frac{\partial}{\partial r}\left(r\frac{\partial u_z}{\partial r}\right)\right). \tag{4.13}$$

Let's start rearranging and integrating.

$$\frac{\partial}{\partial r}\left(r\frac{\partial u_z}{\partial r}\right)=\frac{1}{\mu}\frac{\partial p}{\partial z}r, \tag{4.14}$$

$$\int\frac{\partial}{\partial r}\left(r\frac{\partial u_z}{\partial r}\right)dr=\frac{1}{\mu}\frac{\partial p}{\partial z}\int r\,dr. \tag{4.15}$$

The integration results in:

$$r\frac{\partial u_z}{\partial r}=\frac{1}{\mu}\frac{\partial p}{\partial z}\frac{r^2}{2}+C_1. \tag{4.16}$$

Dividing by r and integrating over the radial direction we yield:

$$\int \frac{\partial u_z}{\partial r} dr = \frac{1}{\mu}\frac{\partial p}{\partial z} \int \frac{r}{2} dr + \int \frac{C_1}{r} dr, \tag{4.17}$$

$$u_z(r) = \frac{1}{4\mu}\frac{\partial p}{\partial z} r^2 + C_1 \ln r + C_2. \tag{4.18}$$

We see that the fluid velocity is dependant on the radial position, the pressure drop and the viscosity. Lastly we identify the two integration constants C_1 and C_2. Setting the radial position equal to zero gives, $\ln(0)C_1 = -\inf C_1$, and to insure a finite velocity at the centreline ($r = 0$) of the pipe C_1 must be zero. Furthermore, assuming no slip conditions at the pipe wall leads to $u_z(R) = 0$ and the constant C_2 is given as:

$$C_2 = -\frac{1}{4\mu}\frac{\partial p}{\partial z} R^2. \tag{4.19}$$

Based on the above assumptions the fluid velocity in the z direction is given as:

$$u_z(r) = \frac{1}{4\mu}\frac{\partial p}{\partial z}\left(r^2 - R^2\right) = -\frac{1}{4\mu}\frac{\partial p}{\partial z}\left(R^2 - r^2\right). \tag{4.20}$$

If the pressure change is positive for a positive position change, the velocity is negative as expected, since fluid flows from high to low pressure. The pressure will decrease along a flow fluid in the flow direction. In the following we make a deviation from force balance where for the fluid velocity we get (4.30) which equals (4.20).

4.2.2 From Force Balance

Let's seek another approach in describing the fluid flow in a pipe. Using the illustration in Fig. 4.3 picturing a cylindrical fluid element in a pipe we set up Newton's Second Law for the fluid element as:

$$m\ddot{z} = p_1 A_z - p_2 A_z + \tau_{rz} A_s, \tag{4.21}$$

$$m\ddot{z} = \pi r^2 (p_1 - p_2) + \tau_{rz} 2r\pi L, \tag{4.22}$$

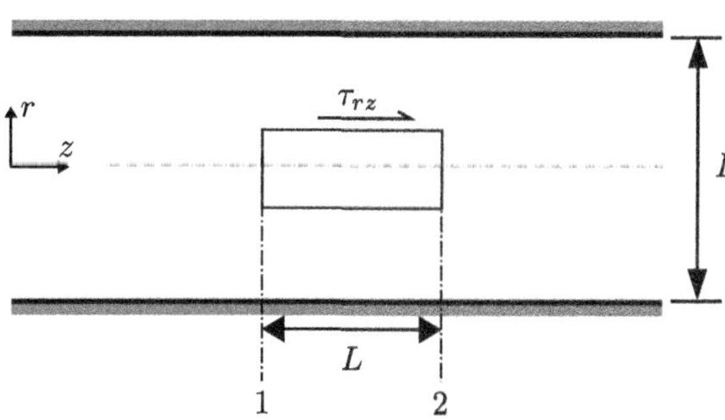

Fig. 4.3 Illustration of a cylindrical fluid element in a tube

where A_z is the circular area of the fluid element in the $r\theta$ plane, while A_s is the surface area of the curved part of the cylindrical fluid element.

Assuming steady flow the acceleration is zero ($\ddot{z} = 0$) yielding:

$$-\tau_{rz} = \frac{r}{2L}(p_1 - p_2). \tag{4.23}$$

Using the definition of shear stresses in a Newtonian fluid, (2.1), the force balance writes,

$$-\mu\frac{d\dot{z}}{dr} = \frac{r}{2L}(p_1 - p_2). \tag{4.24}$$

From this point separation of variables and integration is employed,

$$-d\dot{z} = \tfrac{r}{2\mu L}\Delta p dr, \qquad \Delta p = p_1 - p_2, \tag{4.25}$$

$$-\textstyle\int 1 d\dot{z} = \frac{\Delta p}{2\mu L}\int r dr. \tag{4.26}$$

To derive a function for the fluid velocity as a function of the radial position in the pipe, the integration limits are set to the velocity at the radial position $u(r)$ and the velocity at the wall, which is zero due to no-slip assumption $u(R) = 0$. The radial limits are accordingly set to r and R.

$$-\textstyle\int_{u(r)}^{0} 1 d\dot{z} = \frac{\Delta p}{2\mu L}\int_r^R r dr, \tag{4.27}$$

$$-[u]_{u(r)}^{0} = \tfrac{\Delta p}{2\mu L}\left[\tfrac{r^2}{2}\right]_r^R dr, \tag{4.28}$$

$$-(0 - u(r)) = \tfrac{\Delta p}{2\mu L}\left(\tfrac{R^2}{2} - \tfrac{r^2}{2}\right). \tag{4.29}$$

Hence, the velocity in the pipe across the flow area is given as a function of the radius with maximum speed in the centre $r = 0$ and zero speed at the wall $r = R$. The velocity function as such is given as:

$$u(r) = \frac{\Delta p}{4\mu L}(R^2 - r^2). \tag{4.30}$$

The velocity profile is seen to be parabolic in every section cut in θ. The velocity profile is illustrated in Fig. 4.4.

4.2.3 Volume Flow

The volume flow through the pipe is given as the velocity multiplied by the flow area. However, with the non-constant velocity across the flow area the flow is found by integration as:

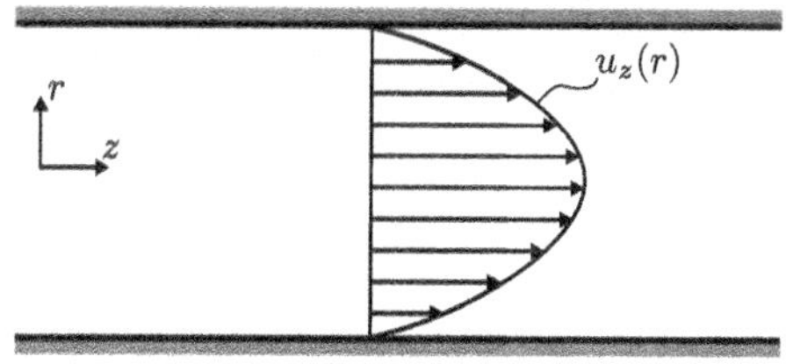

Fig. 4.4 Illustration of fluid velocity profile

$$Q = \int u(r)dA, \tag{4.31}$$

where the area dA is a small ring,

$$dA = 2r\pi dr, \tag{4.32}$$

where the volume flow is given as the definite integral:

$$Q = \int_0^R \frac{\Delta p}{4\mu L}(R^2 - r^2)2r\pi dr, \tag{4.33}$$

$$Q = \frac{\Delta p\pi}{2\mu L}\int_0^R (rR^2 - r^3)dr, \tag{4.34}$$

$$Q = \frac{\Delta p\pi}{2\mu L}\left[\frac{r^2R^2}{2} - \frac{r^4}{4}\right]_0^R = \frac{\Delta p\pi}{2\mu L}\left(\frac{R^4}{2} - \frac{R^4}{4}\right) = \frac{\Delta p\pi}{8\mu L}R^4, \tag{4.35}$$

$$Q = \frac{\Delta p\pi}{128\mu L}D^4. \tag{4.36}$$

In Eq. (4.36) it is seen that the laminar flow through a circular pipe is in linear proportion to the pressure drop. Furthermore the flow is linearly dependent on the inverse of the dynamic viscosity and the length of the tube. It is seen that the diameter of the pipe has a huge influence on the flow-pressure relationship. A small change in diameter leads to a huge change in flow at constant pressure drop since the diameter is in the power 4 in (4.36).

Note that we assumed laminar flow as only viscous forces are included in terms of shearing stresses.

4.2.4 Turbulent Flow in Pipes

Flows in which the inertia forces dominate are said to be turbulent. In turbulent flows the fluid particles do not move in the nice layered pattern as in laminar flow. The particle path is chaotic and the flow is in all spacial directions in the pipe, hence why it is infeasible to seek an analytical solution to the velocity profile or for the laminar flow. The velocity profile is somewhat different from the parabolic for a laminar flow; in turbulent flow the velocity is seen as much more "equal" across the flow cross-section as illustrated in Fig. 4.5.

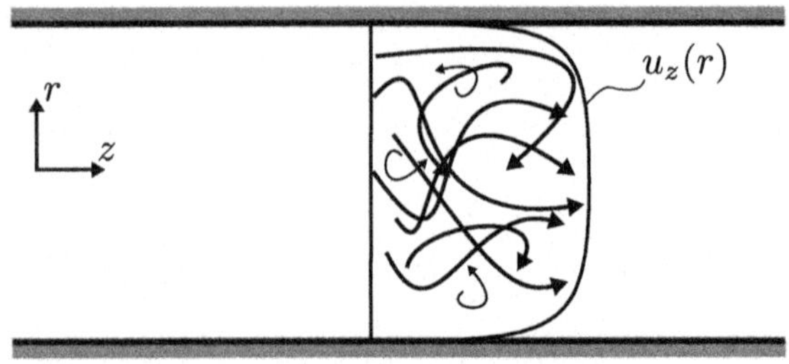

Fig. 4.5 Illustration of velocity profile n a turbulent pipe flow

Darcy's equation (4.37) describes the pressure loss in a fluid line when a fluid flow is present.

$$\Delta p = \lambda \frac{L}{D} \rho \frac{\mathbf{u}^2}{2}, \tag{4.37}$$

where the friction coefficient λ varies with the flow regime and $\mathbf{u}$ is the mean fluid velocity. For a laminar flow regime λ is given as:

$$\lambda = \frac{64}{Re}. \tag{4.38}$$

Note that inserting the laminar friction coefficient of (4.38) in Darcy's equation (4.37) yields the Hagen-Poiseuille equation (4.36). For a turbulent flow regime λ is given as:

$$\lambda = 0.3164 \frac{1}{Re^{\frac{1}{4}}}. \tag{4.39}$$

Note that the friction coefficients introduced here are all for a smooth inner surface of the pipe. For insight on how to account for surface roughness the reader should refer to other textbooks.

4.2.5 Summary of Flow in Pipe

The models presented here are for steady flow so no dynamic change over time is included.

Laminar flow in a circular pipe:

$$Q = \frac{\Delta p \pi}{128 \mu L} D^4, \tag{4.40}$$

$$u_z(r) = \frac{1}{4\mu} \frac{\partial p}{\partial z} \left(r^2 - R^2\right), \tag{4.41}$$

$$u_z(r) = \frac{\Delta p}{4 \mu L} (R^2 - r^2). \tag{4.42}$$

Flow in a circular pipe:

$$\Delta p = \lambda \frac{L}{D} \rho \frac{\mathbf{u}^2}{2}, \tag{4.43}$$

$$\lambda = \frac{64}{Re}, \text{ Laminar flow} \tag{4.44}$$

$$\lambda = 0.3164 \frac{1}{Re^{\frac{1}{4}}}, \text{ Turbulent flow} \tag{4.45}$$

4.3 Flow in Gaps—Leakage Flows

In fluid power systems one often encounters gaps which may be more or less desirable. In a hydraulic cylinder e.g. the clearance between the piston and the cylinder bore in which the piston moves creates a gap. Gaps like these are sometimes sealed with various sealing materials but at other times left unsealed, yielding a hydrostatic bearing as well as yielding a leakage flow. Assuming the surfaces forming the gaps to be parallel, the flow through the gaps may be analytically derived as in the following subsection.

In the case of the gap between the cylinder bore and the piston in a hydraulic cylinder, the gap is parallel if the bore and piston are perfectly circular and concentric, as seen on the illustration in Fig. 4.6. However, the gaps are circular as well.

Let's assume that as the circular gap is unwrapped it almost forms a rectangular gap. If the gap height $h = \frac{D-d}{2}$ is small compared to the gap width $w = \frac{\pi(D+d)}{2}$ assuming it to have a rectangular form is feasible. The gap length here is the height of the piston, L.

4.3.1 *From Force Balance*

In this subsection the flow through a parallel gap is derived from force balance. It is assumed that the flow is steady and that it is laminar so that inertia forces are neglected and only pressure and viscous shear forces are included. No slip conditions are applied at both walls, however no end walls are assumed in the width of the gap.

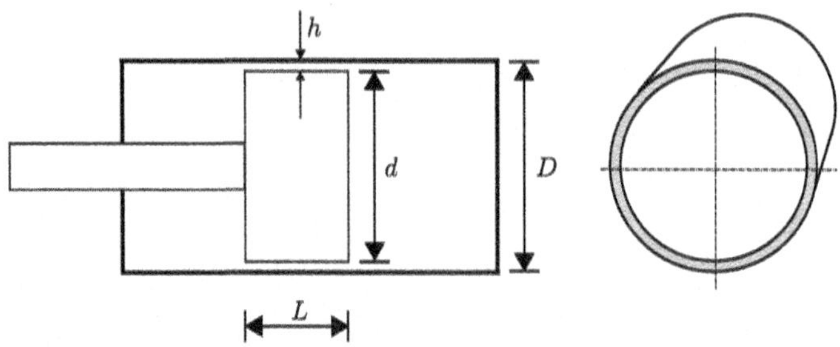

Fig. 4.6 Illustration of gap in fluid power cylinder

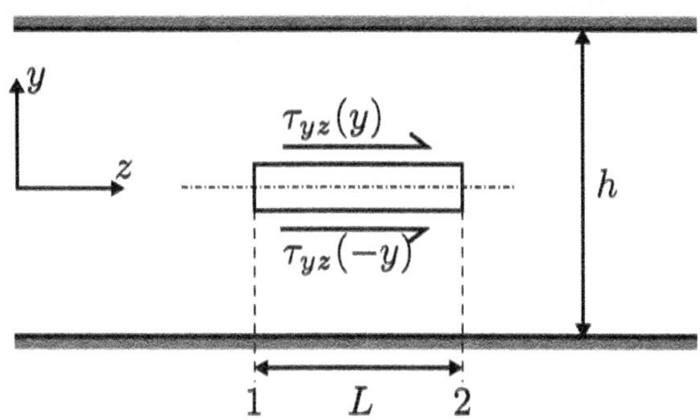

Fig. 4.7 Illustration of a rectangular box fluid element in a flow gap

For the derivation the notation is as illustrated in Fig. 4.7. The flow direction is aligned with the $z-$axis and the gap height is in the $y-$direction. The zx-plane is placed half way between the two walls forming the gap. Applying Newton's Second Law on the fluid element in Fig. 4.7 yields:

$$0 = p_1 2yw - p_2 2yw + \tau_{yz}(y)wL + \tau_{yz}(-y)wL, \tag{4.46}$$

where $2y$ and w are the height and width of the element respectively. The pressure at the ends of the element are p_1 and p_2. Setting the velocity at the two walls equal to zero and placing the fluid element at the centreline justifies the assumption of the shearing stresses on the element surfaces to be equal, $\tau_{yz}(y) = \tau_{yz}(-y)$. Furthermore, using the pressure difference $\Delta p = p_1 - p_2$ yields:

$$0 = 2yw(p_1 - p_2) + 2\tau_{yz}(y)wL = 2yw\Delta p + 2\tau_{yz}(y)wL. \tag{4.47}$$

Including the relation between strain rate and shear stresses for Newtonian fluids, (2.1), yields,

$$0 = 2yw\Delta p + 2wL\mu\frac{d\dot{z}}{dy}. \tag{4.48}$$

The separation of variables and integration brings an expression for the fluid velocity as a function of the y-position in the flow gap:

$$-L\mu\tfrac{d\dot{z}}{dy} = y\Delta p, \tag{4.49}$$

$$-d\dot{z} = \tfrac{\Delta p}{\mu L} y dy, \tag{4.50}$$

$$-\int_{u_z(y)}^{0} d\dot{z} = \tfrac{\Delta p}{\mu L}\int_{y}^{\frac{h}{2}} y dy, \tag{4.51}$$

$$-[\dot{z}]_{u_z(y)}^{0} = \tfrac{\Delta p}{\mu L}\left[\tfrac{y^2}{2}\right]_{y}^{\frac{h}{2}}, \tag{4.52}$$

$$-[0 - u_z(y)] = \tfrac{\Delta p}{\mu L}\left[\tfrac{(\frac{h}{2})^2}{2} - \tfrac{y^2}{2}\right], \tag{4.53}$$

$$u_z(y) = \tfrac{\Delta p}{\mu L}\left(\tfrac{h^2}{8} - \tfrac{y^2}{2}\right), \qquad y \in \left[-\tfrac{h}{2}\ \tfrac{h}{2}\right]. \tag{4.54}$$

Note that the velocity profile is symmetrical around the centre plane zx as $u_z(y) = u_z(-y)$.

4.3.2 *Volume Flow*

The volume flow through the gap is found by integration of the velocity multiplied by the flow area over the hole gap.

$$Q(y) = u_z(y)dA = u_z(y)wdy, \tag{4.55}$$

$$Q = \int u_z(y)dA = \int u_z(y)wdy = \int \tfrac{\Delta p}{\mu L}\left(\tfrac{h^2}{8} - \tfrac{y^2}{2}\right) wdy, \tag{4.56}$$

$$= \tfrac{\Delta pw}{\mu L}\int_{-\frac{h}{2}}^{\frac{h}{2}}\left(\tfrac{h^2}{8} - \tfrac{y^2}{2}\right)dy = \tfrac{\Delta pw}{\mu L}\left[\tfrac{h^2}{8}y - \tfrac{y^3}{6}\right]_{-\frac{h}{2}}^{\frac{h}{2}}, \tag{4.57}$$

$$= \tfrac{\Delta pw}{\mu L}\left(\left[\tfrac{h^2}{8}\tfrac{h}{2} - \tfrac{\frac{h}{2}^3}{6}\right] - \left[\tfrac{h^2}{8}\left(-\tfrac{h}{2}\right) - \tfrac{-\frac{h}{2}^3}{6}\right]\right), \tag{4.58}$$

$$= \tfrac{\Delta pw}{\mu L}\left(\left[\tfrac{h^3}{16} - \tfrac{h^3}{42}\right] - \left[-\tfrac{h^3}{16} + \tfrac{h^3}{42}\right]\right) = \tfrac{\Delta pw}{\mu L}\left(\tfrac{h^3}{8} - \tfrac{h^3}{24}\right), \tag{4.59}$$

$$Q(y) = \tfrac{\Delta pw}{12\mu L}h^3 \tag{4.60}$$

It is seen that the flow-pressure relationship is proportional with some geometric constant (and viscosity) which is expected with a laminar flow regime. Furthermore, it is seen that the gap height has a large influence on the proportionality constant as it contributes in the power of 3.

4.3.3 Velocity Profile from Naiver-Stokes Equation

Naturally the flow through parallel gaps may be derived from Naiver-Stokes Equation. In the current subsection this is shown with the same notation as used above, see Fig. 4.7.

As the flow direction is in the $z-$direction the third direction of the Naiver-Stokes equation for incompressible flow (3.88) is reprinted here for deviation of the velocity profile in the fluid flow gap in Fig. 4.7

$$\rho\left(\frac{\partial u_3}{\partial t}+u_1\frac{\partial u_3}{\partial x_1}+u_2\frac{\partial u_3}{\partial x_2}+u_3\frac{\partial u_3}{\partial x_3}\right)=\rho g_{x_3}-\frac{\partial p}{\partial x_3}+\mu\left(\frac{\partial^2 u_3}{\partial x_1^2}+\frac{\partial^2 u_3}{\partial x_2^2}+\frac{\partial^2 u_3}{\partial x_3^2}\right). \tag{4.61}$$

As a steady flow is assumed, the time dependent acceleration is zero; furthermore the flow in the first and second direction is assumed zero.

$$\frac{\partial u_3}{\partial t}=0, \tag{4.62}$$

$$u_1=u_2=0. \tag{4.63}$$

It is assumed that the flow does not accelerate along the flow direction.

$$\frac{\partial u_3}{\partial x_3}=0, \tag{4.64}$$

$$\frac{\partial^2 u_3}{\partial x_1^2}=0, \tag{4.65}$$

$$\frac{\partial^2 u_3}{\partial x_3^2}=0. \tag{4.66}$$

Hence, this yields a much reduced equation,

$$0=-\frac{\partial p}{\partial x_3}+\mu\frac{\partial^2 u_3}{\partial x_2^2}. \tag{4.67}$$

The notation in Fig. 4.7 leads to $x=x_1, y=x_2, z=x_3$ and $\dot{x}=u_1, \dot{y}=u_2, \dot{z}=u_3$. Integration leads to a function for the velocity by:

$$\frac{\partial p}{\partial z}\int 1dy=\mu\int\frac{\partial^2\dot{z}}{\partial y^2}dy, \tag{4.68}$$

$$\frac{\partial p}{\partial z}(y+C_0)=\mu\frac{\partial\dot{z}}{\partial y}, \tag{4.69}$$

$$\frac{\partial p}{\partial z}\int(y+C_0)dy=\mu\int\frac{\partial\dot{z}}{\partial y}dy, \tag{4.70}$$

$$\frac{\partial p}{\partial z}\left(\frac{y^2}{2}+C_0y+C_1\right)=\mu\dot{z}(y), \tag{4.71}$$

$$\dot{z}(y)=\frac{1}{\mu}\frac{\partial p}{\partial z}\left(\frac{y^2}{2}+C_0y+C_1\right). \tag{4.72}$$

The fluid velocity is here seen to include a second order polynomial in the y coordinate, however the polynomial features two unknown integration constants. The no-slip condition at the walls is the boundary condition utilised to determine the constants C_0 and C_1 as:

$$\dot{z}\left(-\tfrac{h}{2}\right) = \dot{z}\left(\tfrac{h}{2}\right) = 0 \quad \Rightarrow \quad C_0 = 0 \;, \quad C_1 = -\tfrac{h^2}{8}. \tag{4.73}$$

Inserting the constants C_0 and C_1 in Eq. (4.72) results in:

$$\tfrac{\partial p}{\partial z}\left(\tfrac{y^2}{2} - \tfrac{h^2}{8}\right) = \mu \dot{z}(y), \tag{4.74}$$

$$\dot{z}(y) = \tfrac{\Delta p}{\mu L}\left(\tfrac{h^2}{8} - \tfrac{y^2}{2}\right), \qquad \tfrac{\Delta p}{L} = \tfrac{p_1 - p_2}{L} = -\tfrac{\partial p}{\partial z}. \tag{4.75}$$

This velocity profile derived from Navier-Stokes equation is equal to the one in (4.54) derived from force balance, one will yield therefore naturally get the same flow equation, this is left for the reader to show.

4.3.4 Summary on Laminar Flow Between Parallel Plates

Laminar flow between parallel plates.

$$Q = \tfrac{h^3 w}{12\mu L}\Delta p, \tag{4.76}$$

$$u_z(y) = \tfrac{1}{\mu}\tfrac{\partial p}{\partial z}\left(-\tfrac{h^2}{8} + \tfrac{y^2}{2}\right), \qquad y = \left[-\tfrac{h}{2} \; \tfrac{h}{2}\right], \tag{4.77}$$

$$u_z(y) = \tfrac{\Delta p}{\mu L}\left(\tfrac{h^2}{8} - \tfrac{y^2}{2}\right), \qquad y = \left[-\tfrac{h}{2} \; \tfrac{h}{2}\right]. \tag{4.78}$$

Case study–Flow in an Annular

As we shall discuss later the flow paths do not have to be large for a significant flow to appear. The clearance between e.g. a valve spool and valve housing or between cylinder bore and piston may be of a size yielding a significant flow when a large pressure difference is experienced.

Let's assume a clearance between piston and cylinder bore of 40 μm and a piston with a diameter of 80 mm and a length of 30 mm. We assume that the annular around the piston may be unfolded as depicted in Fig. 4.8.

Assuming a VG32 fluid at 40 C° with a density of 886 $\frac{\text{kg}}{\text{m}^3}$ the flow in the annular is given as:

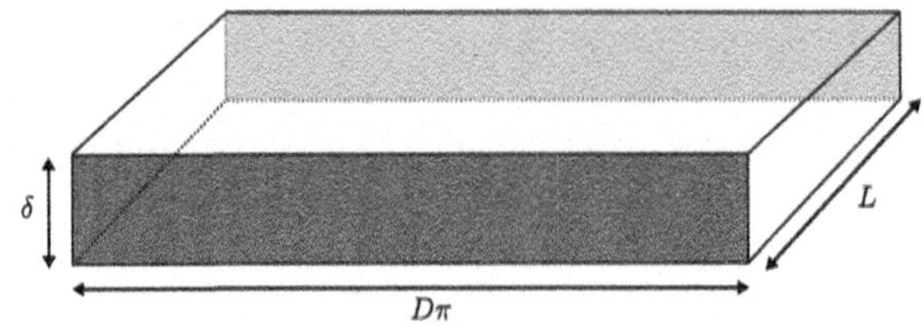

Fig. 4.8 Illustration of the gap in the case study, with the two surfaces marked respectively as in and out flow

Table 4.1 Leakage flow versus pressure difference for the different gap sizes

Q [L/min] μm	Δp [bar]					
	50	100	150	200	250	300
$\delta = 20$	0.06	0.12	0.18	0.24	0.3	0.36
$\delta = 40$	0.47	0.95	1.42	1.89	2.36	2.84
$\delta = 60$	1.56	3.19	4.79	6.38	7.98	9.57

$$Q = \frac{\delta^3 \pi D}{12 \mu L} \Delta p, \tag{4.79}$$

$$Q = \frac{(40 \cdot 10^{-6} \text{m})^3 \pi 0.080 \text{m}}{12 \cdot 32 \cdot 10^{-6} \frac{\text{m}^2}{\text{s}} \cdot 886 \frac{\text{kg}}{\text{m}^3} 0.030 \text{m}} \Delta p. \tag{4.80}$$

For the given gap the flows for various pressure differences are given in Table 4.1. The table further includes flows for a smaller and larger gap.

The flow is seen to increase linearly with the increase in pressure difference, as expected due to the laminar flow regime. Also as expected, the flow change due to gap height is seen to be non-linear in nature; in fact doubling the gap height from 20 μm to 40 μ increases the flow by a factor of 8.

4.4 The Orifice Equation

Orifices are of great importance in fluid power control systems. We shall later discuss how orifices, or restrictions modelled as orifices, are used to control fluid flow and pressure.

An orifice is a sudden restriction of short length in a flow path (ideally, of zero length). As we shall see later the flow area of the orifice may be fixed but may also be variable in some cases. As the orifice has a flow area smaller than the upstream 'pipe' the flow velocity will increase due to continuity and be higher in the orifice. This change in flow velocity will induce a pressure drop for high Reynolds numbers. The pressure drop is due to the acceleration of fluid particles from the upstream velocity

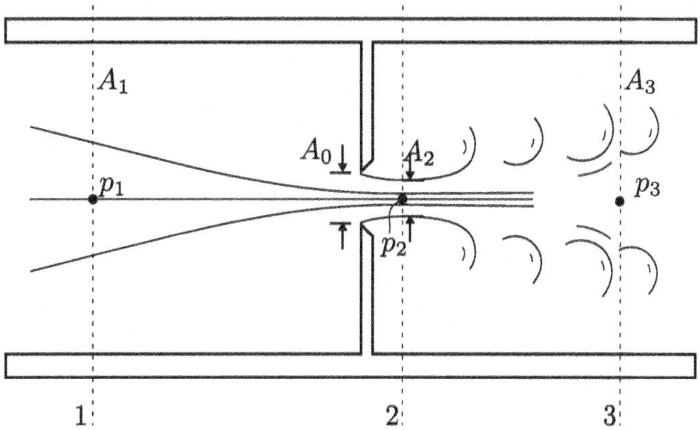

Fig. 4.9 Illustration of an orifice

to the higher jet velocity [1]. For low Reynolds numbers the pressure drop is due to shear forces resulting from the fluid viscosity.

Let's follow a particle through the orifice in Fig. 4.9 and describe it before and after the orifice. Let's consider a constant fluid flow. With Bernoulli's equation we know that the energy along a streamline in an inviscid fluid is constant, hence we may suggest:

$$\frac{p}{\rho} + \frac{u^2}{2} = \text{constant}, \tag{4.81}$$

which may be used for two points along the stream line, such as:

$$\frac{p_1}{\rho} + \frac{u_1^2}{2} = \frac{p_2}{\rho} + \frac{u_2^2}{2}, \tag{4.82}$$

$$u_2^2 - u_1^2 = \frac{2}{\rho}(p_1 - p_2). \tag{4.83}$$

Using the continuity equation for incompressible flow imposes the flow in points 1, 2 and 3 in Fig. 4.10 to be the same such as,

$$Q = A_1 u_1 = A_2 u_2 = A_3 u_3, \tag{4.84}$$

and,

$$u_1 = \frac{A_2}{A_1} u_2. \tag{4.85}$$

Using this formulation of u_1 and Bernoulli's equation, the velocity u_2 in the jet stream may be derived:

$$u_2^2 - \left(\frac{A_2}{A_1} u_2\right)^2 = \frac{2}{\rho}(p_1 - p_2), \tag{4.86}$$

$$u_2^2 \left(1 - \frac{A_2^2}{A_1^2}\right) = \frac{2}{\rho}(p_1 - p_2). \tag{4.87}$$

The mean fluid velocity at point 2 may as such be given as,

$$u_2 = \left(1 - \frac{A_2^2}{A_1^2}\right)^{-\frac{1}{2}} \sqrt{\frac{2}{\rho}(p_1 - p_2)}. \tag{4.88}$$

Multiplying with the flow area to get the volume flow we get:

$$Q = A_2 u_2 = \frac{A_2}{\left(1 - \frac{A_2^2}{A_1^2}\right)^{\frac{1}{2}}} \sqrt{\frac{2}{\rho}(p_1 - p_2)}. \tag{4.89}$$

Note that the flow area in the jet stream, A_2, is not equal to the physical area of the orifice, A_0. Therefore the contraction constant C_c is defined as the ratio between the jet stream and orifice areas as in:

$$C_c = \frac{A_2}{A_0}. \tag{4.90}$$

To simplify the equation and to include minor losses the discharge coefficient C_d is defined as:

$$C_d = \frac{C_v C_c}{\sqrt{1 - \frac{C_c^2 A_0^2}{A_1^2}}}, \tag{4.91}$$

where the coefficient C_v adds some minor losses.

Experimental studies have shown that the static pressure level from point 1 is not recovered in point 3. The pressure in point 3 is seen to be equal to the pressure in point 2. Energy is lost and the pressure in the jet stream may be set equal to the down stream pressure 3. This is why, when employing the orifice equation one uses the pressure before and after the orifice (p_1 and $p_2 = p_3$).

The orifice equation, as we shall later use in modelling fluid power systems and components, is given as:

$$Q = C_d A_0 \sqrt{\frac{2}{\rho} |p_1 - p_2|} \frac{(p_1 - p_2)}{|p_1 - p_2|}. \tag{4.92}$$

Note that the physical opening area is used. Furthermore, the absolute pressure difference is used in the square root while the sign of the pressure difference is multiplied at the end. In other words, the sign of the pressure difference dictates the flow direction, with flow from point 1 to 2 being positive flow direction. An

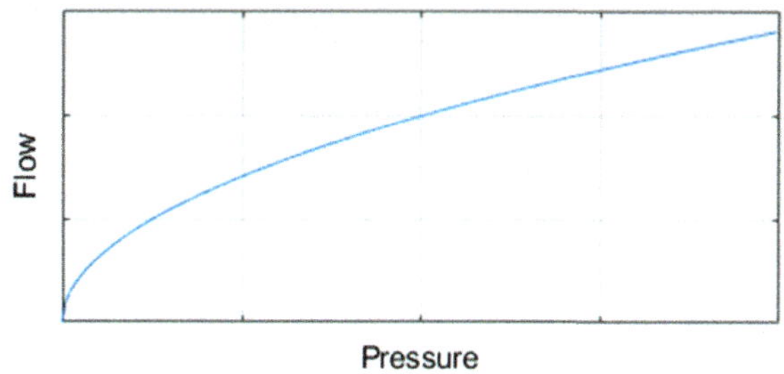

Fig. 4.10 Flow curves for the orifice equation with discharge coefficient and density as $C_d = 0.7$ and $\rho = 883$ kg/m^3

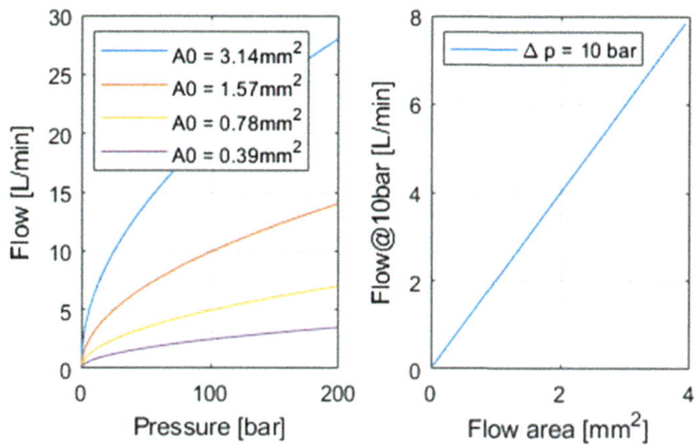

Fig. 4.11 Flow curves for the orifice equation with discharge coefficient and density as $C_d = 0.7$ and $\rho = 883$ kg/m^3

illustration of the relation between pressure and flow is given in Fig. 4.10 in which the non-linear relation is clearly seen.

The pressure flow relation is in Fig. 4.11 shown for four specific orifice areas together with the area flow relation for a constant 10 bar pressure drop.

One should note the linear area flow relation experienced for a constant pressure drop across the orifice.

4.4.1 Laminar Versus Turbulent Orifice Flow

In the derivation of the orifice equation, empirical data suggests that all kinetic energy is lost from the point of vena contractor to point 3 downstream, mainly due to turbulent friction losses. However, for a small flow, the fluid velocity may become

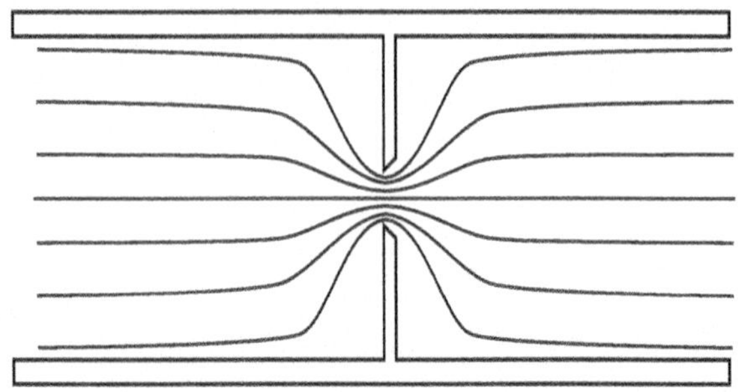

Fig. 4.12 Illustration of laminar orifice flow

low and a laminar flow regime may be experienced through the orifice, as illustrated in Fig. 4.12.

In such a case losses are dominated by the shearing friction related to viscosity. Various means of modelling the laminar flow regime in an orifice exist. In [1] a model using a modified discharge coefficient C_d is presented. The discharge coefficient is in that model a function of the Reynolds number, i.e. the flow velocity, density and viscosity.

The model given in [1] modifying the orifice equation, (4.92), such that it models a laminar flow, is shortly introduced. To distinguish between laminar and turbulent flow Reynolds number is used. For an orifice Reynolds number is given as:

$$\mathrm{Re} = \frac{\rho \frac{Q}{A_0} D_h}{\mu}, \tag{4.93}$$

where the jet velocity is $\frac{Q}{A_0}$ and D_h is the hydraulic diameter. For low Reynolds numbers (Re < 10) investigations have shown the discharge coefficient to be describable by:

$$C_d = \delta \sqrt{\mathrm{Re}}. \tag{4.94}$$

By inserting this modified discharge coefficient in the orifice equation one yields:

$$Q = \delta \sqrt{\frac{\rho \frac{Q}{A_0} D_h}{\mu}} A_0 \sqrt{\frac{2}{\rho}(p_1 - p_2)}, \tag{4.95}$$

$$Q^2 = \delta^2 \frac{\rho \frac{Q}{A_0} D_h}{\mu} A_0^2 \frac{2}{\rho}(p_1 - p_2), \tag{4.96}$$

$$Q = \frac{2\delta^2 D_h A_0}{\mu}(p_1 - p_2), \tag{4.97}$$

where δ is the Laminar flow coefficient which is geometry dependent and often set to 0.2 for sharp edge round orifices.

Reference

1. Merritt H (1967) Hydraulic Control Systems

Part II
Fluid Power Components

Part II introduces the main components of fluid power systems. A brief functional description of each component is followed by the describing equations. For some components multiple model types, e.g. steady state and dynamic models are presented as being applicable to various types of system models which are developed in Part III. The focus of this Lecture Note is on the function of the fluid power systems and not on the physical design and layout of components and systems. The mechanical construction of the various components is not the focus of this note, so the reader is asked to consult other materials on the mechanical and structural design of fluid power components.

In this part we will introduce, e.g. fluid power pumps, motors, cylinders, valves, accumulators, pipes, hoses and fittings. The current part will introduce each component by itself; systems constructed with multiple components are left for Part III where fluid power systems are the focus.

Chapter 5
Fluid Power Pumps

Abstract This chapter introduces a steady state flow model for displacement pumps. Both ideal and non-ideal pump models are introduced. The models are kept on an over level and do not treat internal flow, pressure, and losses. The chapter ends with a brief introduction to various pump topologies, e.g., gear and piston pumps.

Pumps are now and then referred to as the heart of a system. The purpose of the pump is to transform mechanical power to fluid power. Typically a mechanical rotational motion is transformed into movement of a fluid.

Several pump types exist, e.g. velocity pumps (centrifugal), displacement pumps, gravitational pumps. Fluid power systems working with hydrostatic power transmission have displacement pumps as the preferred type. The scope of this lecture note includes various types of displacement pumps; furthermore the control of the pumps is included as well as a brief introduction to discrete displacement pumps.

5.1 Displacement Pumps

The working principal of a displacement pump is, as the name suggests, that the pump displaces a volume of fluid from one place to another. This is done by sucking fluid into a volume by expanding the volume, hence, creating a pressure difference between some reservoir and the expanding volume, yielding an inflow from the reservoir. When the volume is filled with fluid the volume is compressed and oil is pushed (displaced) to another place.

The primary parameter for a displacement pump is the displacement which tells how large the volume displaced every cycle is. We use the symbol D_P for pump displacement in this lecture note.

A. H. Hansen, *Fluid Power Systems*, Fluid Mechanics and Its Applications 129,
https://doi.org/10.1007/978-3-031-15089-0_5

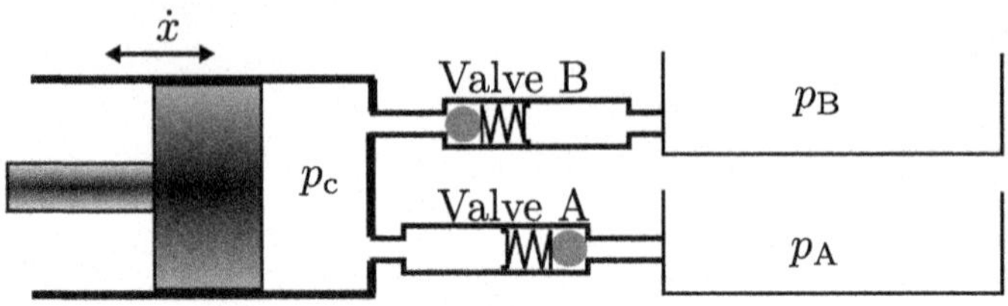

Fig. 5.1 Illustration of a simple one piston displacement pump, using check valves

5.1.1 Single Piston Pump

A simple displacement pump with a single piston pump seen in Fig. 5.1. This pump consists of a cylinder chamber, a piston and two check valves.

The pump illustrated displaces fluid from reservoir A to B. The working principle of the single piston pump is outlined in six steps:

1. The piston moves left and the chamber volume is increased. With no fluid flowing into the chamber the continuity equations state that the pressure will decrease.
2. When the pressure p_c in the chamber is lower than the pressure p_A in reservoir A, valve A will open and allow fluid from reservoir A to flow into the chamber.
3. When the piston movement is stopped the pressure in the chamber and reservoir A will be close to equal, and valve A will close due to the soft spring.
4. The piston now moves right. With no fluid flowing out of the chamber the pressure p_c will increase.
5. When the pressure p_c in the chamber is higher than the pressure p_B in reservoir B, valve B will open and allow fluid to flow from the chamber into reservoir B.
6. When the piston movement is stopped the pressure in the chamber and reservoir B will be close to equal and valve B will close due to the soft spring. And a new cycle can start.

This illustrates the working principle of a displacement pump, where fluid is sucked into the pump at one side and displaced to the other side.

5.2 The General Pump Model—Steady State

When discussing a general displacement pump the focus is on modelling the behaviour of the pump in terms of output flow without addressing the specific type and construction of the pump.

The diagram symbols for fluid pumps are given in Fig. 5.2. The symbols give information on pump functions as:

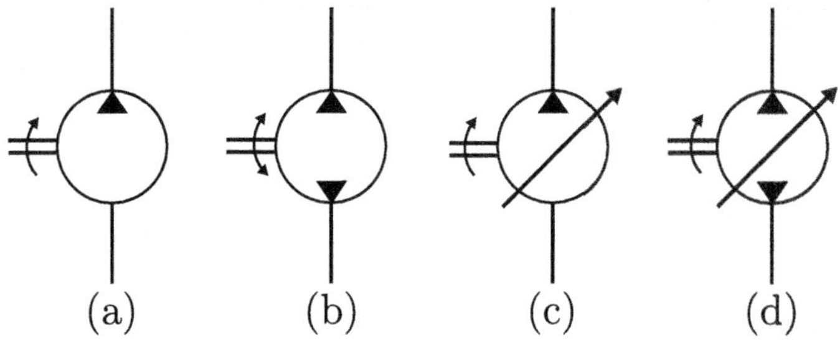

Fig. 5.2 Pump diagram symbols

- Uni- or bi-direction rotation.
- Uni- or bi-direction fluid flow.
- Fixed or variable displacement.

The circle with an arrowhead inside pointing outward indicates a pump delivering fluid in the direction of the arrow head. Pump (a) and (b) are said to have fixed displacement, that is, the displacement volume is constant. Contrary for pump (c) and (d) the displacement volume may be changed. Pump (b) and (d) are able to supply fluid in both directions; for the fixed displacement pump (b) this can be obtained by reversing the pump rotation. The variable pump (d) can also reverse the flow, however, as we shall discuss later the reverse flow direction may also be realised by changing the displacement.

5.2.1 *Ideal Pump Model*

The flow out of an ideal pump is given by the pump displacement and the pump velocity. When using rotational pumps the ideal pump flow is given as:

$$Q_{\mathrm{P}} = D_{\omega}\omega_{\mathrm{P}}, \tag{5.1}$$

where pump displacement D_{ω} $[\frac{m^3}{rad}]$ is the volume displaced per radian of rotation and $\omega_{\mathrm{P}}[\frac{rad}{s}]$ is the rotational velocity of the pump. As mentioned the displacement may be variable for some pump types. Let's assume the displacement to depend on the parameter γ with the pump flow given as:

$$Q_{\mathrm{P}} = D_{\omega}(\gamma)\omega_{\mathrm{P}}. \tag{5.2}$$

Assuming an ideal pump, the mechanical power into the pump $\mathcal{P}_{\mathrm{mP}}$ and the fluid power out of the pump $\mathcal{P}_{\mathrm{fP}}$ is equal, which yields:

$$\mathcal{P}_{\mathrm{mP}} = \mathcal{P}_{\mathrm{fP}}, \tag{5.3}$$

with the mechanical power,

$$\mathcal{P}_{\mathrm{mP}} = T_{\mathrm{P}}\omega_{\mathrm{P}}, \tag{5.4}$$

and the fluid power,

$$\mathcal{P}_{\mathrm{fP}} = Q_{\mathrm{P}}\Delta p_{\mathrm{P}} = D_{\omega}\omega_{\mathrm{P}}\Delta p_{\mathrm{P}}. \tag{5.5}$$

Substituting (5.4) and (5.5) into (5.3) yields:

$$T_{\mathrm{P}}\omega_{\mathrm{P}} = D_{\omega}\omega_{\mathrm{P}}\Delta p_{\mathrm{P}}, \tag{5.6}$$

$$T_{\mathrm{P}} = D_{\omega}\Delta p_{\mathrm{P}}, \tag{5.7}$$

where T_{P} is the torque mechanically applied to the pump and Δp_{P} is the pressure difference from fluid inlet to fluid outlet. It is therefore given that the torque requirement on the machine driving the pump is dependant on the pressure difference across the pump and the pump displacement.

5.2.2 *Non-ideal Pump Model*

As we shall discuss in later sections several pump constructions exist, each with pros and cons. No pump is ideal, however. The designer has to deal with the real life pump and not the ideal mathematical model. Real pumps have some hydro-static lubrications which are designed to decrease mechanical friction in the pump. The hydro-static lubrication, on the other hand, results in leakage flow and viscous friction.

Volumetric Efficiency

In a real pump the displaced volume of fluid is not equal to the theoretical volume expected in (5.1). Some fluid will leak from the output port to the inlet port. For now we accept this leakage flow to be proportional to the pressure difference across the pump and as such given as:

$$Q_{\mathrm{le}} = C_{\mathrm{le}}\Delta p_{\mathrm{P}}, \tag{5.8}$$

where C_{le} is named the leakage coefficient for the pump. With this description the leakage flow is assumed laminar, as is seen from the linear relation between flow and pressure difference.

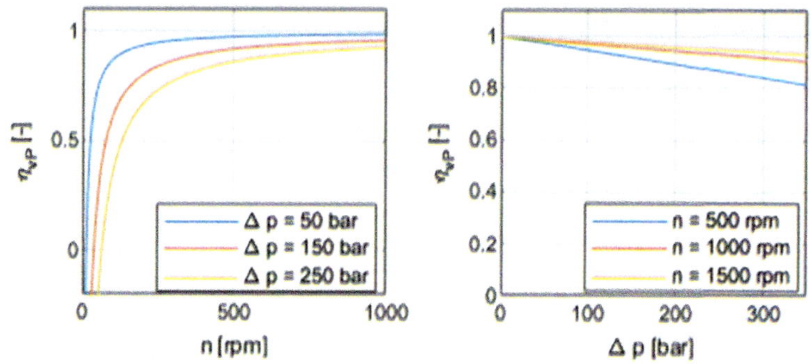

Fig. 5.3 Illustration of how the volumetric efficiency for a pump varies with rotation speed and pressure difference

With this laminar leakage flow the actual flow delivered by the pump is given as the theoretical flow minus the leakage flow,

$$Q_{\mathrm{P}} = Q_{\mathrm{tP}} - Q_{\mathrm{le}}. \tag{5.9}$$

Defining the volumetric efficiency η_{vP} the actual pump flow may be given as:

$$Q_{\mathrm{P}} = Q_{\mathrm{tP}}\eta_{\mathrm{vP}}. \tag{5.10}$$

The volumetric efficiency, η_{vP}, may be given as the ratio between actual and theoretical flow, as:

$$\eta_{\mathrm{vP}} = \frac{Q_{\mathrm{P}}}{Q_{\mathrm{tP}}} = \frac{Q_{\mathrm{tP}} - Q_{\mathrm{le}}}{Q_{\mathrm{tP}}}, \tag{5.11}$$

where Q_{P} is actual pump flow while Q_{tP} and Q_{le} is theoretical pump flow and leakage flow respectively.

The volumetric efficiency is illustrated in Fig. 5.3 as a function of the pump rotational velocity and the pressure difference respectively. Note the negative efficiency for low rotational velocity if a pressure difference is present. This is seen because the leakage flow is larger than the theoretical pump flow. This is, however, not practically possible and only present as the pressure after the pump is assumed constant.

Hydro-mechanical Efficiency

As for all other mechanical systems frictional forces are present in fluid power pumps. This entails that the torque required to drive non-ideal (or real) pumps at a given velocity is higher than the theoretically required torque given by (5.7). As friction phenomenons depends on geometry, surface roughness and movement of the parts, it is complex to model friction prices and especially in a general sense covering a wide range of pump types. However, some general friction terms may be given as:

$$T_{\mathrm{mP}} = C_{\mathrm{mP}} \Delta p_{\mathrm{P}}, \tag{5.12}$$

$$T_{\mathrm{vP}} = C_{\mathrm{vP}} \mu \omega_{\mathrm{P}}, \tag{5.13}$$

$$T_{\mathrm{hP}} = C_{\mathrm{hP}} \omega_{\mathrm{P}}^2, \tag{5.14}$$

$$T_{\mathrm{sP}} = C_{\mathrm{sP}} \omega_{\mathrm{P}}, \tag{5.15}$$

where T_{mP}, T_{vP}, T_{hP} and T_{sP} are mechanical, viscous, turbulent and static torque losses in the pump respectively. For calculation of these the mechanical, viscous, turbulent and static friction coefficient given as C_{mP}, C_{vP}, C_{hP} and C_{sP} respectively are used.

The machine driving the pump must overcome not only the torque due to the pressure raise across the pump, but also these four frictional torques. The actual required torque is then given as:

$$T_{\mathrm{P}} = T_{\mathrm{tP}} + T_{\mathrm{mP}} + T_{\mathrm{vP}} + T_{\mathrm{hP}} + T_{\mathrm{sP}}. \tag{5.16}$$

The hydro-mechanical efficiency is defined as the ratio between theoretical and actual required torque to drive the pump:

$$\eta_{\mathrm{hmP}} = \frac{T_{\mathrm{tP}}}{T_{\mathrm{P}}} = \frac{T_{\mathrm{tP}}}{T_{\mathrm{tP}} + T_{\mathrm{mP}} + T_{\mathrm{vP}} + T_{\mathrm{hP}} + T_{\mathrm{sP}}}, \tag{5.17}$$

where the theoretical torque T_{tP} is given in (5.7) and the actual required torque T_{P} is given in (5.16).

The hydro-mechanical efficiency is illustrated in Fig. 5.4 as functions of the rotational velocity and the pressure difference respectively.

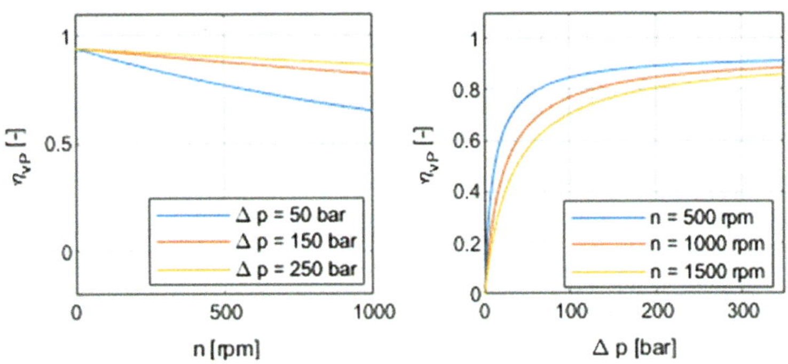

Fig. 5.4 Illustration of hydro-mechanical efficiency of a pump

A simple model for the actual torque required to drive the pump is given as:

$$T_P = \frac{D_\omega \Delta p_P}{\eta_{hmP}} \tag{5.18}$$

5.2.3 Summary on General Pump Model

When modelling pumps the important parameters are the displacement, D_ω, and the rotational velocity ω_P together with the volumetric and hydro-mechanical efficiencies, η_{vP}, η_{hmP}.

Note that the general models discussed in this section do not deal with the dynamic behaviour of the pump; the models in this chapter assume steady state conditions. Neither pressure dynamics in the pump nor mechanical dynamics of the rotational mechanical parts are modelled.

When talking about pumps one must remember that pumps are flow providers, displacement providers. Pumps move a given amount of fluid–they do not set the pressure after the pump, the pressure after the pump is established as the fluid flows through various restrictions or into confined control volumes of fixed or variable size.

Equations for an ideal pump:

$$Q_{tP} = D_\omega(\gamma)\omega_P, \tag{5.19}$$

$$T_{tP} = D_\omega(\gamma)\Delta p_P. \tag{5.20}$$

$$\mathcal{P}_{tfP} = Q_P \Delta p_P, \tag{5.21}$$

$$\mathcal{P}_{tmP} = T_{tP}\omega_P. \tag{5.22}$$

Equations for a non-ideal pump:

$$Q_P = D_\omega(\gamma)\omega_P \eta_{vP}, \tag{5.23}$$

$$T_P = \frac{D_\omega(\gamma)\Delta p_P}{\eta_{hmP}}. \tag{5.24}$$

$$\mathcal{P}_{fP} = Q_{Pt}\Delta p_P \eta_{vP} = D_\omega(\gamma)\omega_P \Delta p_P \eta_{vP}, \tag{5.25}$$

$$\mathcal{P}_{mP} = T_P \omega_P = \frac{D_\omega(\gamma)\Delta p_P}{\eta_{hmP}}\omega_P, \tag{5.26}$$

$$\eta_P = \frac{\mathcal{P}_{fP}}{\mathcal{P}_{mP}} = \eta_{vP}\eta_{hmP}. \tag{5.27}$$

5.3 Pump Types

This section will introduce some basic pump types and illustrate their working principles. The intention is to leave the reader with a visual understanding and idea of how displacement pumps work and give a brief understanding of pump types, but not to illustrate or describe numerous pump designs. Readers seeking a deep understanding of pump design are asked to consult a dedicated textbook.

All pumps addressed in this textbook are, as mentioned, displacement pumps. Pumps may be grouped by their way of displacing fluid from inlet to outlet. The various pump types have varying pros and cons, e.g. pressure level, friction, complexity.

5.3.1 Gear Pumps

Gear type pumps most often consist of two gears interacting as they rotate. In external gear pumps, the two gears or the gear set are placed in the housing as seen left in Fig. 5.5. One of the gears, (the driver) is placed on the driving shaft to which a primary mover delivers a torque. The second gear (the slave) is driven by the driver gear such that input torque is solely delivered to the driving gear. The gears and the housing together form small volumes separated by the gear teeth. Right in Fig. 5.5 two of these small displacement volumes are marked in red.

As the gear rotates, the volumes formed by the housing and the gear teeth are moved from the inlet side along the pump housing to the outlet side, where the volume "collapses" by the interacting teeth. The volumes are again formed as they are moved to the inlet side meaning they may suck in fluid and carry it along the house to the outlet.

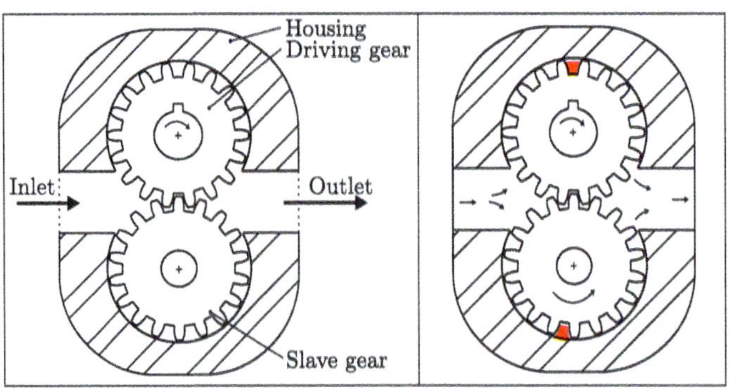

Fig. 5.5 Illustration of an external gear pump

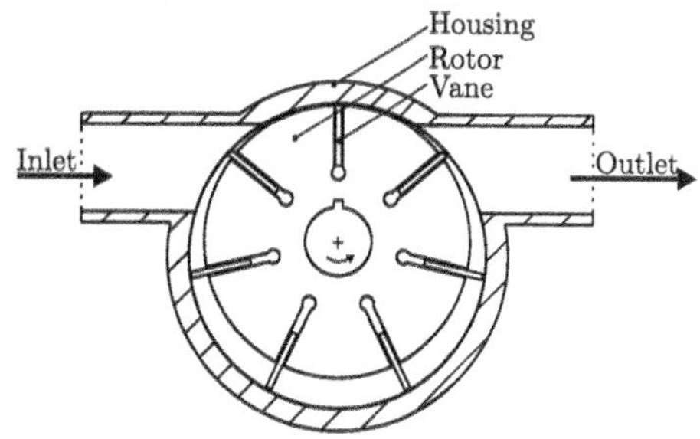

Fig. 5.6 Illustration of a vane pump

The displacement of the gear pump is given by the volumes formed by the teeth and the housing. The pump designer may vary the pump displacement by depth of the gears and the width of the gears in the axial direction, hence into the plane of the drawing in Fig. 5.5. The displacement of gear pumps is most often fixed, hence the fluid flow may be controlled solely with the rotational velocity of the gears.

5.3.2 Vane Pumps

Vane type pumps basically consist of a rotor and a housing. However, the rotor features a driving shaft and a number of vanes. The vanes are mounted radially as seen in Fig. 5.6.

The vanes are pushed outward against the housing either by springs, fluid pressure or both. As in the gear pumps the housing and the vanes form small volumes which are moved along the housing as the rotor rotates. The displacement may be controlled by the eccentricity of the rotor compared to the housing. If the eccentricity is zero the displacement is zero as the volume displaced from inlet to outlet is equal to the volume displaced from outlet to inlet. If the rotor may be moved to both a positive and negative eccentricity, the pump may pump bidirectionally for unidirectional rotor velocity. Hence, with a variable displacement the fluid flow is controlled with both displacement and rotational velocity of the rotor.

5.3.3 Piston Pumps

Piston type pumps may be divided into two major categories; axial and radial pumps.

In the radial pumps the pistons are mounted radially to the driving shaft which turns an eccentric, see Fig. 5.7. The volume in the cylinder bores are hence expanded and compressed as the pistons move according to the movement of the eccentric. This change in volume in the cylinder bores is what performs the displacement of fluid. The connection of the cylinder bores to low and high pressure, inlet and outlet, may be performed in various ways. Check valves may be used as shown in Fig. 5.7.

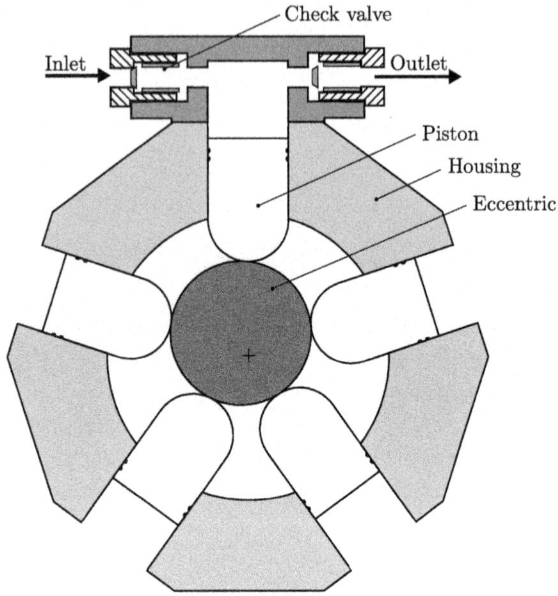

Fig. 5.7 Illustration of a radial piston pump

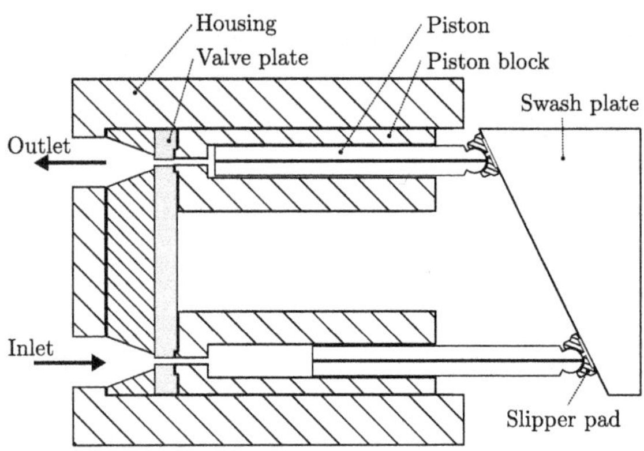

Fig. 5.8 Illustration of axial piston pump with fixed swash plate angle

However, a commutator unit which changes the connection with the rotation of the drive shaft may also be used.

In axial piston pumps the pistons move parallel to the drive shaft. Hence, an eccentric can not perform the transformation from rotation of the shaft to translation of the pistons. Axial piston pumps often consist of a number of pistons moving in a single piston block. The pistons are each supported in a slipper pad standing on a non-rotating swash plate, see Fig. 5.8. The cylinder block, pistons and slipper pads are all rotating with the drive axle. The bore in the cylinder block is connected to inlet and outlet through a non-rotating valve plate.

The translational movement of the pistons is performed by the movement of the slipper pad on the inclining swash plate. As the piston is moved towards the valve plate the current bore is connected to the high pressure outlet displacing fluid to the outlet. When the piston is moving away from the valve plate, the bore is connected to the low pressure inlet, sucking fluid into the bore.

For piston pumps the displacement is set by the number and size of pistons, along with the length of the pulsating movement (the stroke length). In variable displacement pumps the inclination of the swash plate or the eccentricity of the eccentric is varied to control the displacement.

Note that zero eccentricity or zero inclination yields zero displacement, and displacement direction may change if the eccentricity or inclination angle may be negative.

5.3.4 Discrete Displacement Pumps

Discrete Displacement Pumps (DDPs) are constructed as piston pumps. However, instead of check valves, valve plates or commutators, the connection to low pressure inlet and high pressure outlet is performed with active or semi-active valves. Two 2/2 way on/off valves are connected to each piston chamber and to low and high pressure respectively. By controlling the on/off valves the chamber connection to inlet and outlet is controlled. This imposes the opportunity to vary the displacement by connecting a varying number of chambers.

Let's assume that all chambers but one are connected to low pressure; the last is alternating between low and high pressure and pumping fluid. The displacement volume of the full pump is equal to the displacement of the single piston chamber. By using more of the other chambers the displacement of the full pump may be increased. However, the change in displacement is rather discrete, as a discrete number of the chambers are used. The displacement of the pump may be given as:

$$D_{\text{Pump}} = D_{\text{piston}} N \;, \quad N = 1, 2 \,..\, M, \tag{5.28}$$

where the displacement of each piston chamber is given as D_{piston} and N is the number of active pistons, with M being the number of pistons in the pump.

The discrete step between possible flow outputs may be decreased by allowing a part stroke operation, however this encounters various difficulties.

Chapter 6
Rotary Actuator—Motors

Abstract This chapter introduces a steady state flow model for displacement motors. Both ideal and non-ideal motors models are introduced. The models are kept on an over level and do not treat internal flow, pressure, and losses.

Fluid power motors are the rotary actuator of fluid power systems. Motors and pumps are very alike, often pumps and motors may be run in both pumping and motoring mode. However, this imposes extra requirements on the sealing of the machine and possibly induces the need for a drain connection. The diagram symbol for fluid power motors is seen in Fig. 6.1. As with the pumps, motors come with both fixed and variable displacement and uni- and bi-directional flow.

It can be seen that motors (a) and (b) are fixed displacement motors, hence, for a constant flow a constant velocity is expected. By contrast, the motors (c) and (d) are variable displacement motors. Further note, that the motors (a) and (c) allow only unidirectional flow, contrary to (b) and (d) which allow bidirectional flow.

The physical construction of motors is similar to the pump construction, as well as the models. The main difference between pumps and motors is the direction of power transfer. While pumps transfer mechanical power to fluid power, motors transfer fluid power to rotary mechanical power.

6.1 Motor Models

6.1.1 Ideal Motor Model

The ideal motor is equivalent to a pump but with power flow in the opposite direction, such that power flow is from the fluid to a mechanical motion. Fluid flow is here the input, leading to a theoretical rotational velocity ω_{tM} of,

$$\omega_{\mathrm{tM}} = \frac{Q_{\mathrm{M}}}{D_{\omega}}, \tag{6.1}$$

A. H. Hansen, *Fluid Power Systems*, Fluid Mechanics and Its Applications 129,
https://doi.org/10.1007/978-3-031-15089-0_6

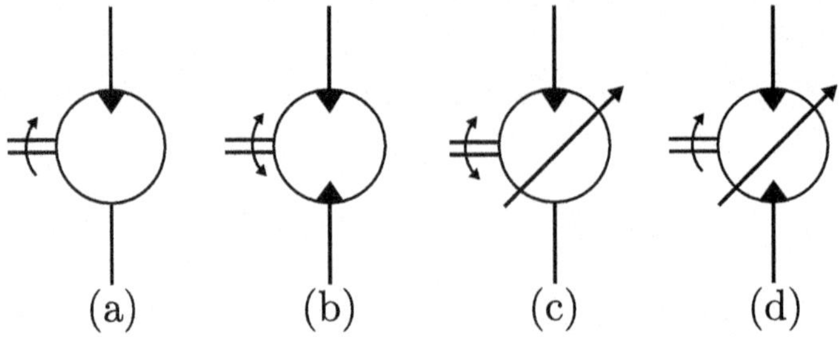

Fig. 6.1 Motor symbols

as with the pump, a variable displacement unit may be used so the model reads,

$$\omega_{\mathrm{tM}} = \frac{Q_{\mathrm{M}}(\gamma)}{D_{\omega}}, \tag{6.2}$$

where D_ω is the displacement and Q_{M} is the fluid volume flow. The torque produced by the motor is given as,

$$T_{\mathrm{tM}} = \Delta p_{\mathrm{M}} D_{\omega}, \tag{6.3}$$

where Δp_{M} is the pressure difference between in- and outlet.

6.1.2 *Non-ideal Motor Model*

Volumetric Efficiency

As with pumps, motors feature losses. Again we shall define a volumetric and hydro-mechanical efficiency. Let's again assume that some leakage is experienced in the motor which may be modelled as,

$$Q_{\mathrm{le}} = C_{\mathrm{le}} \Delta p_{\mathrm{M}}, \tag{6.4}$$

where C_{le} is the coefficient used to model the laminar leakage flow. The motor is driven by the actual flow Q_{M} however some of this flow is not displaced through the motor displacement chambers but is leaking 'around' the motor. The flow performing work is therefore the flow supplied subtracted the leakage flow such that the actual motor velocity becomes,

$$\omega_{\mathrm{M}} = \frac{Q_{\mathrm{M}} - Q_{\mathrm{le}}}{D_{\omega}}, \tag{6.5}$$

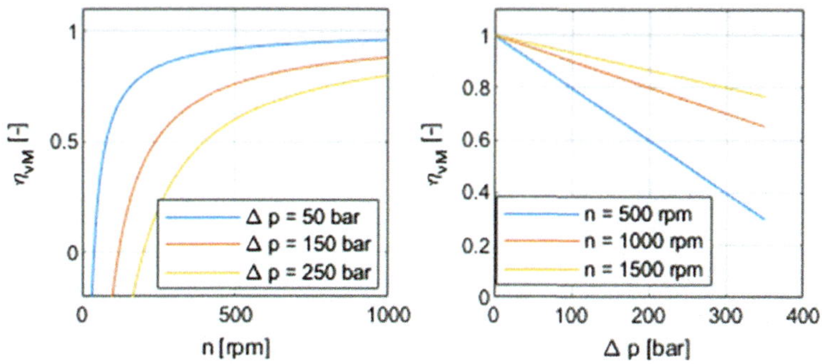

Fig. 6.2 Illustration of pressure and velocity dependency on the volumetric efficiency of a fluid power motor

and the volumetric efficiency η_{vM} is defined as,

$$\eta_{vM} = \frac{\omega_M}{\omega_{tM}} = \frac{\frac{Q_M - Q_{le}}{D_\omega}}{\frac{Q_M}{D_\omega}}, \tag{6.6}$$

where ω_M and ω_{tM} are the actual and theoretical motor velocities respectively while Q_{le} is the motor leakage flow.

The variation in volumetric efficiency for varying rotational velocity and pressure difference is given in Fig. 6.2. Again the simple assumption of constant pressure drop for all rotational velocities imposes an issue in the low velocity area with this very simple model.

Hydro—Mechanical Efficiency

The torque produced on the output shaft of the motor is due to both mechanical and fluid friction being somewhat lower that the theoretical expectation. Friction phenomenons are dependent on geometry, surface roughness and movement of the parts, which makes it complex to model friction; especially in a general sense, covering a wide range of motor types. However, some general friction terms may be given as,

$$T_{mM} = C_{mM} \Delta p_M, \tag{6.7}$$

$$T_{vM} = C_{vM} \mu \omega_M, \tag{6.8}$$

$$T_{hM} = C_{hM} \omega_M^2, \tag{6.9}$$

$$T_{sM} = C_{sM} \omega_M, \tag{6.10}$$

where T_{mM}, T_{vM}, T_{hM} and T_{sM} are mechanical, viscous, turbulent and static torques respectively. For calculation of the mechanical, viscous, turbulent and static torques, the coefficients given as C_{mM}, C_{vM}, C_{hM} and C_{sM} respectively are used.

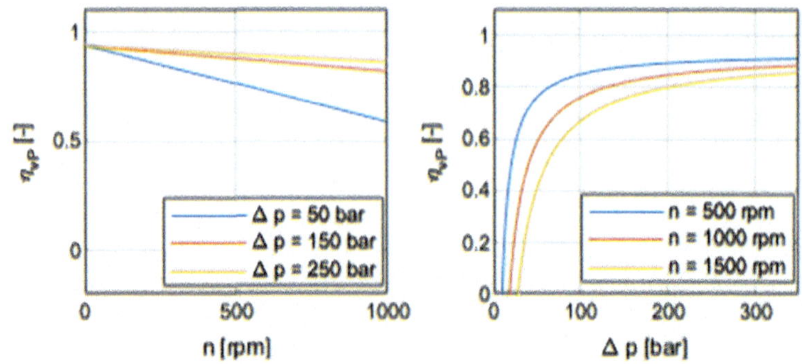

Fig. 6.3 Illustration of pressure and velocity dependency on the hydro-mechanical efficiency of a fluid power motor

The produced torque is the theoretical torque subtracted these friction torques:

$$T_{\mathrm{M}} = T_{\mathrm{tM}} - (T_{\mathrm{mM}} + T_{\mathrm{vM}} + T_{\mathrm{hM}} + T_{\mathrm{sM}}), \tag{6.11}$$

and the hydro-mechanical efficiency, η_{hmM}, is defined as,

$$\eta_{\mathrm{hmM}} = \frac{T_{\mathrm{M}}}{T_{\mathrm{tM}}}, \tag{6.12}$$

where T_{M} and T_{tM} are the actual and theoretical torques from the motor respectively.

Figure 6.3 illustrates how the hydro-mechanical efficiency varies with pressure difference across the motor and with the rotational velocity of the motor.

Chapter 7
Linear Actuators—Cylinders

Abstract This chapter introduces the linear actuator of fluid power systems. A lumped parameter time domain model is derived for a "standard" differential cylinder, including a simple friction model based on viscous and coulomb terms. The model is derived based on the flow continuity equation, Newtons second law and a laminar leakage flow model. The time domain model is reduced to a steady state model usable during a system design phase. Lastly, the concept of Multi-Chamber Cylinders and their modelling is introduced.

In fluid power systems linear motion is typically performed using a cylinder. In this chapter we will derive models for fluid power cylinders, both for differential and symmetrical cylinders as well as multi-chamber cylinders. The scope of the chapter is on modelling the cylinder pressure, piston motion and force. System analysis for linear actuation systems is addressed in Part III.

A fluid power cylinder typically consists of a cylinder barrel, a piston and a piston rod, see Fig. 7.1. Seals is placed in the piston seal and the end caps sealing between the two chambers and each chamber and the exterior respectively.

7.1 Differential Cylinder

A diagram symbol of a differential cylinder is seen in Fig. 7.2 together with an illustration of the piston areas. The name differential entails that the piston areas are unequal, the effect of which we shall discuss later. The cylinder dimensions are often given as D/d/L, with the piston diameter named D and the rod diameter named d while the stroke length is given as L. One may note that the piston in the illustration is not able to move the full stroke length and stay inside the cylinder end caps. It is therefore important to distinguish between diagram symbols, illustration with notation, and technical drawings. However, as this type of notation with "too short" cylinder bore is common we shall also use it here.

A. H. Hansen, *Fluid Power Systems*, Fluid Mechanics and Its Applications 129,
https://doi.org/10.1007/978-3-031-15089-0_7

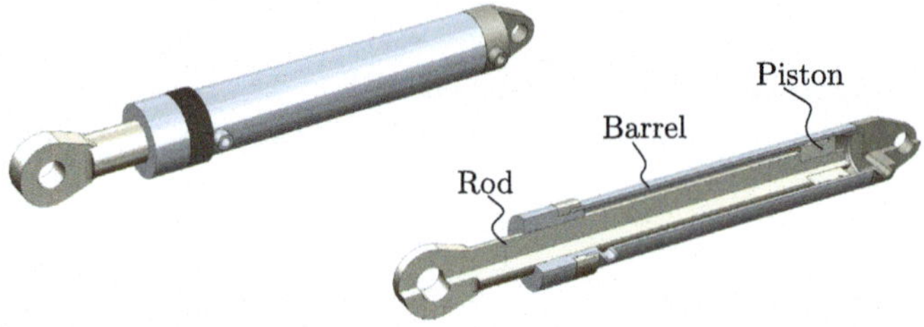

Fig. 7.1 Illustration of a fluid power cylinder

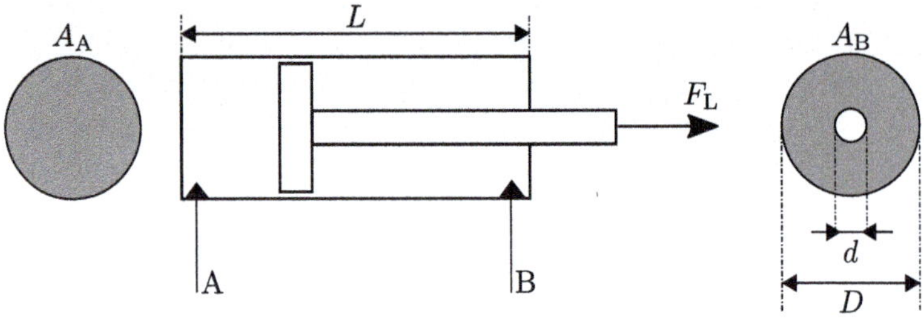

Fig. 7.2 Illustration of a fluid power cylinder

The piston areas shown on Fig. 7.2 are given as:

$$A_A = D^2 \frac{\pi}{4}, \tag{7.1}$$

and,

$$A_B = D^2 \frac{\pi}{4} - d^2 \frac{\pi}{4} = (D^2 - d^2)\frac{\pi}{4}. \tag{7.2}$$

The piston area ratio is defined as:

$$\alpha = \frac{A_B}{A_A} \tag{7.3}$$

The area ratio may vary from zero to one as the larger piston area is named A_A. An illustration of cylinders featuring a cylinder ratio of zero and one are seen in Fig. 7.3.

Cylinder (a) is a one chamber cylinder where the B side is vented to air and a spring is applied to produce a return force. The (b) cylinder is a symmetric cylinder where the piston rod goes through the whole cylinder bore yielding equal piston areas for the A and B side.

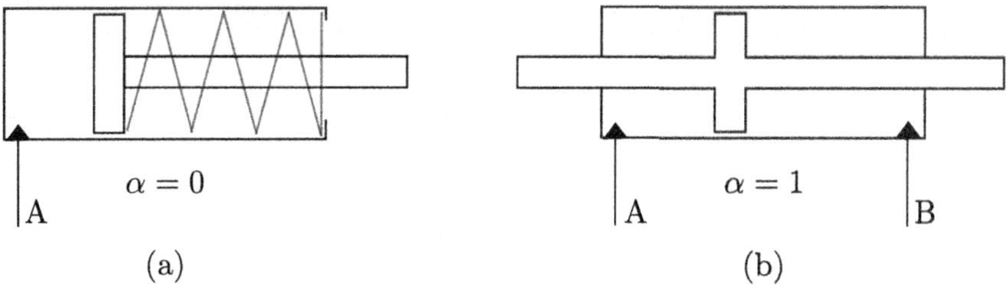

Fig. 7.3 Illustration of piston area ratio limits

7.1.1 Modelling

When modelling a cylinder we utilised the continuity equation to derive differential equations including the chamber pressures. The flow into the piston side chamber A and the flow out of the rod side chamber B are often seen as inputs to the cylinder model. Whereas, the leakage flows Q_{le} and $Q_{\mathrm{le.ext}}$ are a part of the cylinder model in which a laminar flow model is adopted for the leakage flows. Piston movement is modelled using Newton's Second Law.

The notation seen in Fig. 7.4 is utilised throughout the modelling. One may note that the flow direction of the volume flow Q_{A} and Q_{B} is such that positive flow corresponds to positive piston velocity. This common sign convention may ease the evaluation of time domain simulation as well as steady state calculations. The pressure in the two chambers are p_{A} and p_{B}, Q_{le} is the leakage flow from A to B while the leakage from the A and B chambers are $Q_{\mathrm{le.extA}}$ and $Q_{\mathrm{le.extB}}$ respectively.

Continuity Equation

The basic continuity equation states:

$$Q_{\mathrm{in}} - Q_{\mathrm{out}} = \dot{V} + \tfrac{V}{\beta}\dot{p}. \tag{7.4}$$

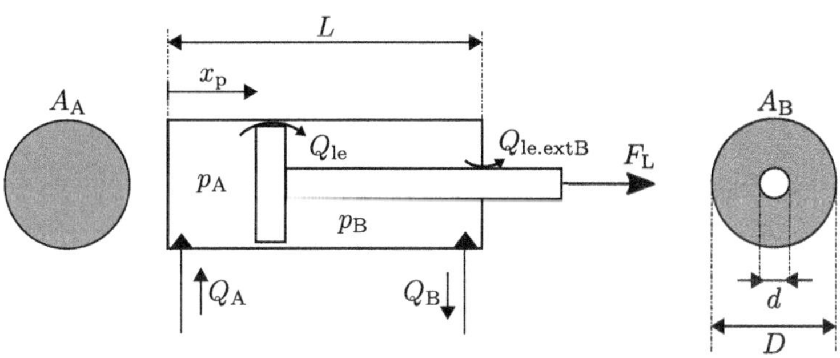

Fig. 7.4 Illustration for modelling of a differential cylinder

Employing the continuity equation for chambers A and B yields:

$$A: \quad Q_{\mathrm{A}} - Q_{\mathrm{le}} - Q_{\mathrm{le.extA}} = A_{\mathrm{A}}\dot{x}_{\mathrm{p}} + \frac{V_{\mathrm{A}}(x_{\mathrm{p}})}{\beta}\dot{p}_{\mathrm{A}}, \tag{7.5}$$

$$B: \quad Q_{\mathrm{le}} - Q_{\mathrm{le.extB}} - Q_{\mathrm{B}} = -A_{\mathrm{B}}\dot{x}_{\mathrm{p}} + \frac{V_{\mathrm{B}}(x_{\mathrm{p}})}{\beta}\dot{p}_{\mathrm{B}}. \tag{7.6}$$

Let's dwell on the nature of the signs for the various terms. In (7.5) the flow into the chamber is Q_{A} therefore this is added, whereas the leakage flow from chamber A to chamber B is outflow as well as the external leakage from chamber A (note that $Q_{le.extA}$ is included for the generality of the model even when not present in Fig. 7.4). Q_{le} and $Q_{le.extA}$ are therefore subtracted. On the right hand side the volume expansion is positive, with positive piston velocity. For the B chamber the inflow is only the leakage flow from the A chamber. The flow Q_{B} and the external leakage from chamber B are both outflow and as such, subtracted. The volume in the B chamber is contracting for a positive piston velocity, thereby the term $A_{\mathrm{B}}\dot{x}_{\mathrm{p}}$ is subtracted.

Let's assume a laminar flow regime for the leakage flows (we expect small gap heights) where we express the leakage with a proportionality constant named the leakage coefficient.

$$Q_{\mathrm{le}} = C_{\mathrm{le}}(p_{\mathrm{A}} - p_{\mathrm{B}}), \tag{7.7}$$

$$Q_{\mathrm{le.extA}} = C_{\mathrm{le.A}}\,p_{\mathrm{A}}, \tag{7.8}$$

$$Q_{\mathrm{le.extB}} = C_{\mathrm{le.B}}\,p_{\mathrm{B}}. \tag{7.9}$$

Let's define the chamber volume functions as:

$$V_{\mathrm{A}}(x_{\mathrm{p}}) = V_{\mathrm{Ad}} + A_{\mathrm{A}}x_{\mathrm{p}}, \tag{7.10}$$

$$V_{\mathrm{B}}(x_{\mathrm{p}}) = V_{\mathrm{Bd}} + A_{\mathrm{B}}\left(L - x_{\mathrm{p}}\right), \tag{7.11}$$

where V_{Ad} and V_{Bd} are the dead volume of chambers A and B respectively and L is the stroke length. Rewriting the continuity equation such that all terms including the pressure gradient are on the left hand side yields:

$$\frac{V_{\mathrm{A}}(x_{\mathrm{p}})}{\beta}\dot{p}_{\mathrm{A}} = Q_{\mathrm{A}} - C_{\mathrm{le}}(p_{\mathrm{A}} - p_{\mathrm{B}}) - C_{\mathrm{le.A}}\,p_{\mathrm{A}} - A_{\mathrm{A}}\dot{x}_{\mathrm{p}}, \tag{7.12}$$

$$\frac{V_{\mathrm{B}}(x_{\mathrm{p}})}{\beta}\dot{p}_{\mathrm{B}} = C_{\mathrm{le}}(p_{\mathrm{A}} - p_{\mathrm{B}}) - C_{\mathrm{le.B}}\,p_{\mathrm{B}} - Q_{\mathrm{B}} + A_{\mathrm{B}}\dot{x}_{\mathrm{p}}, \tag{7.13}$$

which is the proper form ready to be solved numerically.

Newton's Second Law

The movement of the piston may be modelled using Newton II. A free body diagram for the combined elements of piston and piston rod is seen in Fig. 7.5. In the top of the figure, pressure in each chamber is shown as the distributed loads they are, however in the bottom of the figure the pressure loads are summed to forces working in the centre of the load distributions.

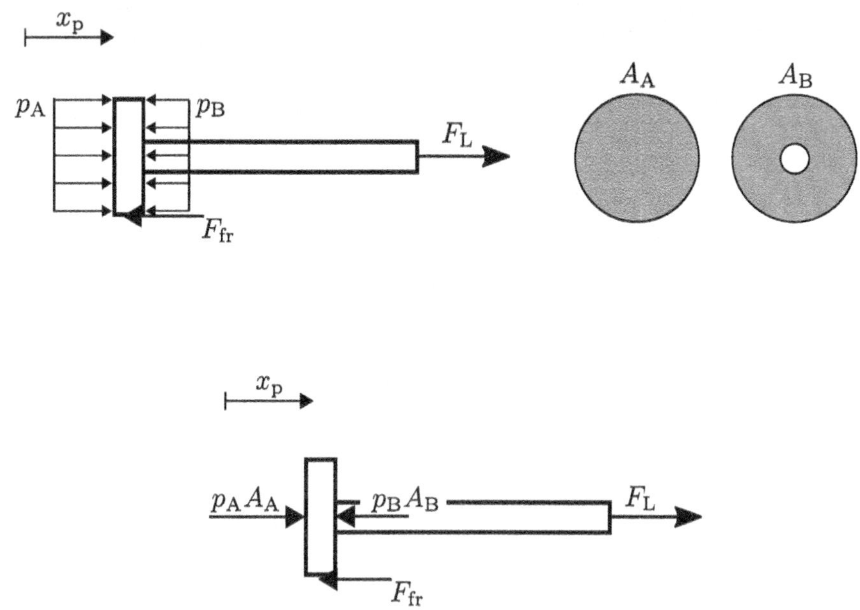

Fig. 7.5 FBD of the piston

Newton II for the piston is given as:

$$M\ddot{x}_p = p_A A_A - p_B A_B - F_{fr}(t) + F_L(t). \tag{7.14}$$

Assuming that the friction force may be modelled as a coulomb and a viscous friction yields:

$$F_{fr}(t) = F_c sgn(\dot{x}_p) + B_p \dot{x}_p, \tag{7.15}$$

where F_c is the constant coulomb friction, while B_p is the viscous friction coefficient. Inserting this friction model and rewriting to the form of a non-homogeneous second order differential equation yields:

$$M\ddot{x}_p + B_p \dot{x}_p = p_A A_A - p_B A_B - F_c sgn(\dot{x}_p) + F_L(t). \tag{7.16}$$

Note the rather simple friction model. Evidently numerous friction models exist and may be utilised in the modelling of a fluid power cylinder. However, in this note we settle for a simple model and encourage the reader to seek further information on friction models elsewhere.

7.1.2 Steady State Model

In the design phase, steady state modelling is often utilised i.e., the system response without the transient phase is used as a design base when laying out the fluid power system. For the ideal fluid power cylinder, the leakage flow and the friction is assumed zero. As such, the dynamic model of the ideal cylinder is:

$$\frac{V_A(x_p)}{\beta}\dot{p}_A = Q_A - A_A\dot{x}_p, \tag{7.17}$$

$$\frac{V_B(x_p)}{\beta}\dot{p}_B = -Q_B + A_B\dot{x}_p, \tag{7.18}$$

$$M\ddot{x}_p + B_p\dot{x}_p = p_A A_A - p_B A_B - F_L(t). \tag{7.19}$$

Cutting away the transient response means that the pressure gradient and the piston acceleration are zero, such that, the chamber pressures and the piston velocity are constant in time, for a constant load force.

Constructing the ideal model for a fluid power cylinder in steady state, then, gives,

$$Q_A = A_A\dot{x}_p, \tag{7.20}$$

$$Q_B = A_B\dot{x}_p, \tag{7.21}$$

$$F_{tC}(t) = p_A A_A - p_A A_B, \tag{7.22}$$

where the theoretical cylinder force is $F_{tC}(t)$. Naturally, no ideal fluid power cylinders exist therefore the hydro-mechanical and the volumetric efficiency are introduced. The hydro-mechanical efficiency gives the ratio between expected force from the cylinder chamber pressure on the load and the actual force transferred to the load.

$$\eta_{hmC} = \frac{F_{aC}}{F_{tC}} = \frac{F_{tC} - F_{fr}}{F_{tC}} = \frac{(p_A A_A - p_B A_B) - F_{fr}}{p_A A_A - p_B A_B}. \tag{7.23}$$

The actual piston force on the load is given as:

$$F_{aC} = \eta_{hmC}\left(p_A A_A - p_B A_B\right). \tag{7.24}$$

The volumetric efficiency is often set to one due to cylinder sealings being effective.

Power

The power delivered from a fluid power cylinder to the load, or the mechanism driven by the cylinder, is given as force multiplied by velocity:

$$\mathcal{P} = \dot{x}_P F_C, \tag{7.25}$$

where the force given from the chamber pressure and the velocity is found based on the fluid flow, such that the power is:

$$\mathcal{P} = \tfrac{Q_A}{A_A}\eta_{hmC} A_A (p_A - \alpha p_B), \tag{7.26}$$

$$\mathcal{P} = \eta_{hmC} Q_A (p_A - \alpha p_B), \tag{7.27}$$

$$\mathcal{P} = \eta_{hmC} (Q_A p_A - Q_B p_B). \tag{7.28}$$

Note that in cases of positive piston velocity, the fluid power supplied to the cylinder is $Q_A p_A$ whereas $Q_B p_B$ is fluid power delivered "from" the cylinder. The power exerted on the load is hence the difference in fluid power multiplied by the efficiency.

7.1.3 Summary

A two chamber cylinder has been modelled, based on the continuity equation, Newton's Second Law and the flow equation for laminar flow in small gaps. Both steady state equations and equations including the transient response have been derived. Note that the port flow into and out of the cylinder chambers are seen as inputs to the model. We will later combine the cylinder models with valve or pump models, which models the port flow.

General two chamber cylinder model:

$$\tfrac{V_A(x_p)}{\beta}\dot{p}_A = Q_A - Q_{le} - Q_{le.extA} - A_A \dot{x}_p, \tag{7.29}$$

$$\tfrac{V_B(x_p)}{\beta}\dot{p}_B = Q_{le} - Q_{le.extB} - Q_B + A_B \dot{x}_p, \tag{7.30}$$

$$M\ddot{x}_p = p_A A_A - p_B A_B - F_{fr}(t) - F(t), \tag{7.31}$$

with leakage flows,

$$Q_{le} = C_{le}(p_A - p_B), \tag{7.32}$$

$$Q_{le.extA} = C_{le.A} p_A, \tag{7.33}$$

$$Q_{le.extB} = C_{le.B} p_B, \tag{7.34}$$

and chamber volumes,

$$V_A(x_p) = V_{A0} + A_A x_p, \tag{7.35}$$

$$V_B(x_p) = V_{B0} - A_B x_p, \tag{7.36}$$

and friction model,

$$F_{fr}(t) = F_c \tanh\left(\frac{\dot{x}_p}{\gamma}\right) + B_p \dot{x}_p. \tag{7.37}$$

Steady state model:

$$F_{aC} = \eta_{hmC}\,(p_A A_A - p_A A_B) \tag{7.38}$$

$$\dot{x}_p = \frac{Q_A}{A_A}, \qquad \dot{x}_p = \frac{Q_B}{A_B} \tag{7.39}$$

In the steady state equations the hydro-mechanical and the volumetric efficiencies are employed to account for non-ideal cylinders.

Finally, the reader should realise that the model derived may be utilised both for differential and symmetric cylinders with A_A and A_B being equal for a symmetric cylinder.

7.2 Multi-chamber Cylinder

A multi-chamber cylinder is a cylinder with more than two working chambers. Multi-chamber cylinders may be specially constructed as a cylinder with more than two chambers or e.g. by two standard cylinders mechanically linked together, see Fig. 7.6. On the left of the figure, two standard differential cylinders are mechanically connected to form a four-chamber cylinder; a dedicated three-chamber cylinder is illustrated on the right.

The model of a multi-chamber cylinder is much like the standard two chamber one, though differing in the number of continuity equations and pressure dependent force terms. The general model for a multi-chamber cylinder is given in shorthand notation, where N is the number of cylinder chambers and i is a cylinder chamber counter, hence an integer running from 1 to N. The model includes Newton's Second Law, N continuity equations and a friction model:

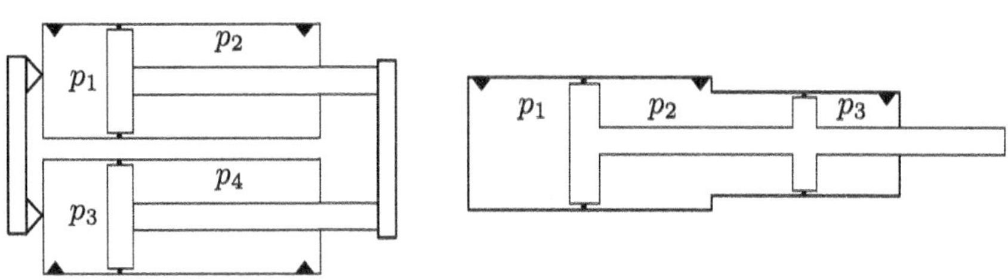

Fig. 7.6 Illustration of multi-chamber cylinders

$$M\ddot{x}_{\mathrm{p}} = \Sigma_{i=1}^{N} p_i A_i - F_{\mathrm{fr}}(t) - F_{\mathrm{L}}(t), \tag{7.40}$$

$$\dot{p}_i = \frac{\beta(p_i)}{V_i(x_{\mathrm{p}})}\left(Q_i - A_i\dot{x}_{\mathrm{p}}\right), \tag{7.41}$$

$$F_{\mathrm{fr}}(t) = F_{\mathrm{c}}\mathrm{sgn}(\dot{x}_{\mathrm{p}}) + B_{\mathrm{p}}\dot{x}_{\mathrm{p}}, \tag{7.42}$$

where A_i and p_i are the piston area and the pressure in the i'th cylinder chamber respectively. Note that the cylinder areas must include a sign handling the force direction, i.e. if the pressure p_2 leads to a force in a negative direction, the area A_2 must be negative. In the current model, the flow Q_i is always the flow into the i'th chamber, hence outflows are negative. The creator of the model is, however, free to chose a sign convention for piston velocity and flow. Another nice approach would be to have positive flows for a positive piston velocity, at least in steady state.

In addition to the momentum and continuity equations a number of equations for leakage flow could be included in the model where chamber inflow and load force are model inputs.

Chapter 8
Control Elements—Valves

Abstract This chapter introduces various valves that are used to control, e.g., flow direction, flow amount or pressure level at various points in a fluid power system. A flow model for each valve is setup from the orifice equation and Newton's second law modelling the moving parts in each valve. Piloted valves request a model including some continuity equations describing pilot pressures. Throughout the chapter several illustrations of industry valves are included. Valves introduced are include check valve, on/off valve, needle throttle valve, pressure compensated flow control valve, proportional valve, servo valve, pressure compensated proportional valve, pressure relief valve, pressure reduction valve, pressure control valve.

Valves are an important control element in fluid power systems. Various valves control direction and amount of flow throughout a fluid power system. Specific valves control the flow so that a preset pressure level is obtained in certain pressure nodes.

In this chapter we shall discuss various type of valve, both in terms of their geometric structure and their functionality. Furthermore, the actuation of the valves will briefly be treated. We will develop models with various complexities for the valves, as we discuss in later chapters when to employ the various models.

Valves may be characterised in several ways, e.g. by the function they perform (directional, pressure, flow etc.), the way they are built into the system (inline, sandwich, cartridge etc.), the way they restrict flow (ball, poppet, spool etc.), and so on. By the function of the valves we shall in this lecture note familiarise ourselves with "Directional" valves, "Pressure control" valves, "Flow control" valves and "Proportional" valves.

A. H. Hansen, *Fluid Power Systems*, Fluid Mechanics and Its Applications 129,
https://doi.org/10.1007/978-3-031-15089-0_8

8.1 General Valve Models

Valves function by restricting/controlling the flow area from port to port. The flow through most valves may be modelled with the orifice equation as:

$$Q = C_d A_0(x_v)\sqrt{\frac{2}{\rho}|p_1 - p_2|}\frac{p_1 - p_2}{|p_1 - p_2|}, \tag{8.1}$$

where the opening area A_0 is a function of the valve spool/poppet position x_v. The pressure p_1 is upstream from the valve and p_2 is the downstream pressure from the valve. x_v is fluid density while C_d is the discharge coefficient which depends on the valve geometry.

The movement of the valve spool/poppet is modelled with Newton's Second Law as:

$$\ddot{x}_v M_v = A_1 p_1 + A_2 p_2 + A_{p.i} p_{p.i} - F_{fr}(\dot{x}_v) + F_{fl}(t) + F_{spring}(x_v) + F_{act}(t), \tag{8.2}$$

where A_1 and A_2 are the areas on which the pressure before and after the valve works. $A_{p.i}$ is a number of areas on which pilot pressure $p_{p.i}$ works. All areas must include the sign for the force direction. Hence, the moving part of the valve is exposed to the pressure before and after the valve. Further, it can be actuated by multiple pilot pressures. In addition to the pressure driven forces, the moving part of the valve is, as all other moving parts, affected by friction, given as $F_{fr}(\dot{x}_v)$. When fluid flows through the valve the flow may change direction multiple times inside the valve. These directional changes induce momentum change which may impose a flow force, $F_{fl}(t)$, on the valve. The last two forces in (8.2) are a spring and an actuation force.

Depending on the valve construction the moving valve part may be exposed to some of the forces mentioned above; but as we shall see later, not all are necessarily present.

8.2 Directional Valves

Directional valves are used to control the flow direction by opening and closing various flow paths in a fluid power system. Some directional valves control multiple flow paths simultaneously, while others control only a single flow path. The special case of proportional directional valves may control both flow direction and flow amount by proportionally controlling the flow area.

As seat, ball, poppet and spool type valves are the most used we shall familiarise with these type of directional valves. Though several valve functions may be performed with various means of restricting the flow, only one or two ways are shown for each valve function in this lecture note.

8.2.1 Check Valve

The function of a check valve is to allow flow in only one direction in a fluid flow path. Hence, this valve is e.g. used to protect pumps against back flow. The diagram symbols for three types of check valves are depicted in Fig. 8.1.

First, the figure shows a basic check valve (a), where fluid will flow left to right (A→B) as long as pressure at port A is higher than pressure at port B. Conversely, the valve is closed and allows no flow if the B pressure is higher than the A pressure. The second check valve (b) is a spring-loaded check valve allowing flow from port A to B when the pressure difference $\Delta p = p_A - p_B$ is sufficient to overcome the spring force. The opening pressure for the spring-loaded check valve is often in the range of 0.5 bar to 5 bar. Flow from port B to A is also not allowed for the spring-loaded check valve (b). The piloted check valve (c) may on the other hand allow flow from port B to A if a sufficient pilot pressure is imposed—the pilot connection is shown with the dashed line. If no pilot pressure is imposed the valve works as a normal check valve (a).

Check valves are often ball, poppet or plate type. Figure 8.2 illustrates how a ball type check valve functions. The A pressure works on the flow area projected onto the ball; this will also be the effective area for the B pressure.

The ball movement in the check valve may be modelled with Newton's Second Law as:

$$M\ddot{x}_v = d^2\frac{\pi}{4}(p_A - p_B) - \dot{x}_v B_v - x_v k_s, \tag{8.3}$$

where M is the mass of the ball, k_s and B_v are the spring constant and damping coefficient respectively. The ball position x_v is dependent on the pressure at port A

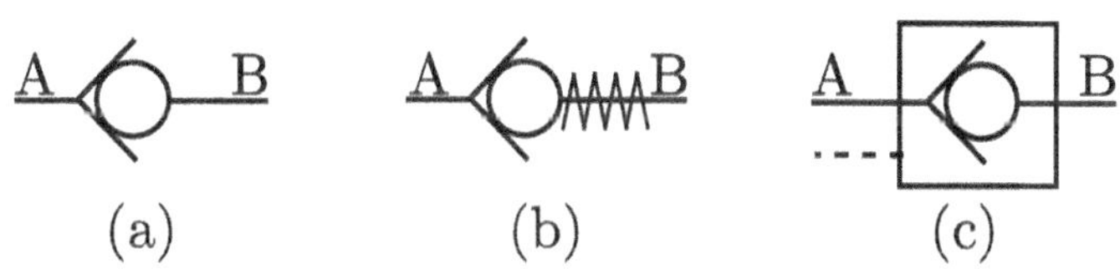

Fig. 8.1 Diagram symbols for check valves. **a** Check valve, **b** Spring-loaded check valve, **c** Piloted check valve

Fig. 8.2 Illustration of a spring-loaded check valve

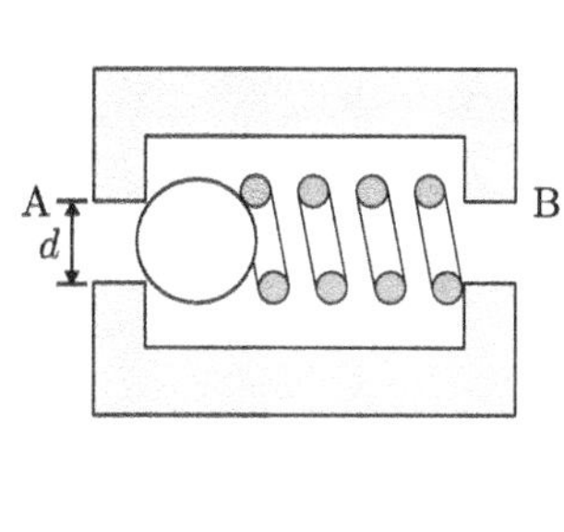

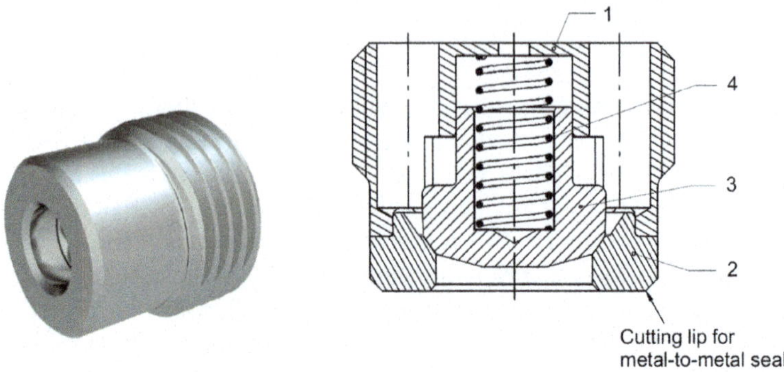

(a) (b)

Fig. 8.3 3D illustration of a Bucher Hydraulics check valve. RKVE ...-VD (**a**). Section view from the valve data sheet (**b**) [1]

and B (p_A and p_B) and the size of the flow area at A, $d^2\frac{\pi}{4}$. The flow through a check valve is modelled with the orifice equation:

$$Q = C_d A_0(x_v)\sqrt{\frac{2}{\rho}(p_A - p_B)} \tag{8.4}$$

Note the dependency on ball position which again depends on pressure drop.

Example of Industry Valve

The Bucher Hydraulics RKVE..-VD is an industrial check valve. It is a spherical poppet-type designed as a screw-in valve for block mounting. A 3D and cut through view is seen in Fig. 8.3. Figure 8.3b taken from the valve data sheet shows the valve bode, 1, valve seat, 2, valve poppet, 3, and the valve spring 4.

The diagram symbol used in the data sheet is seen in Fig. 8.4 together with flow curves for the size 10 valve, with three spring configurations for opening pressure.

8.2.2 On-Off Valves

The simplest active controlled directional valve is the on/off valve or 2/2 way valve. As the name suggests it is a valve that is either on or off, i.e., it lets flow pass or blocks all flow in the current direction. Fluid power diagram symbols for some on/off valves are seen in Fig. 8.5. The valves are seen to have two states: a and b, on and off.

Note the variety in the symbols for the closed state. The ones including the symbol of a check valve indicate that the valve is drop tight in the current direction. A spring dictates the normal (de-energised) state of the valve as no actuation force is applied.

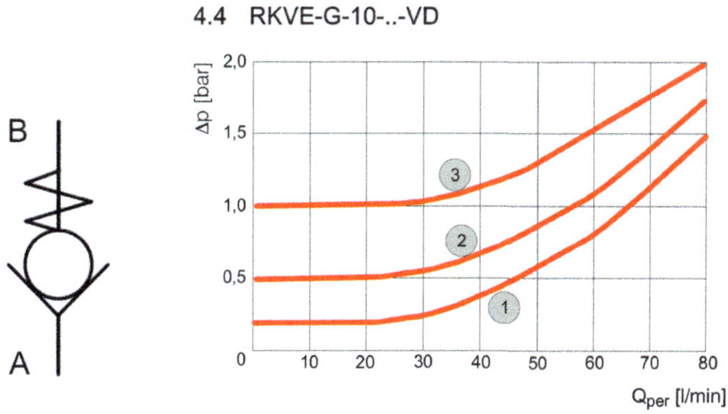

(a) (b)

Fig. 8.4 Diagram symbol of a Bucher Hydraulics check valve. RKVE …-VD (**a**). Flow pressure curves from the valve data sheet (**b**)[1]

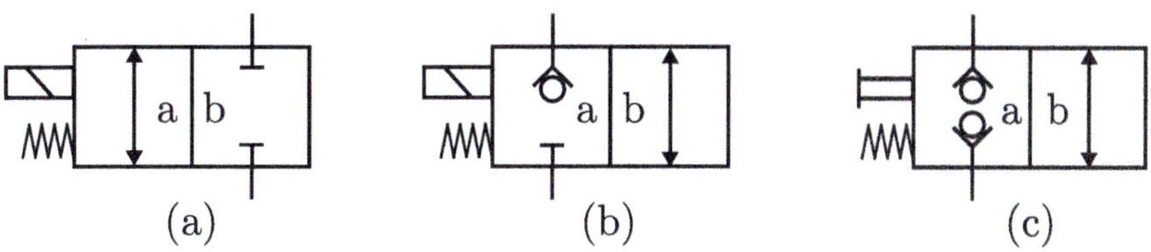

Fig. 8.5 Diagram symbols for On/off valves

Example of Industry Valve

The Bucher hydraulics WS22NG5 is a direct acting solenoid actuated poppet type on/off valve. As indicated in Fig. 8.6b both the normally closed or open valve is virtually leakage free in both directions. A 3D drawing is seen in Fig. 8.6a to give a sense of the physical construction.

A part cut-through view and flow curve from the valve data sheet is given in Fig. 8.7. It is seen that the orifice equation models flow through the valve rather well.

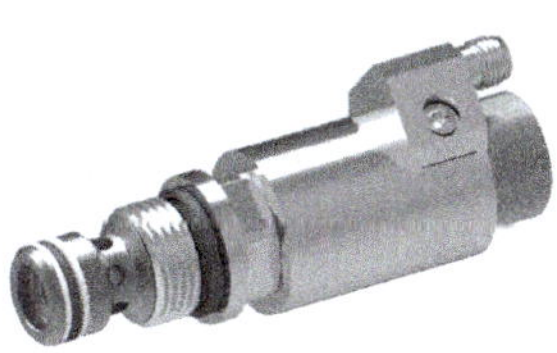

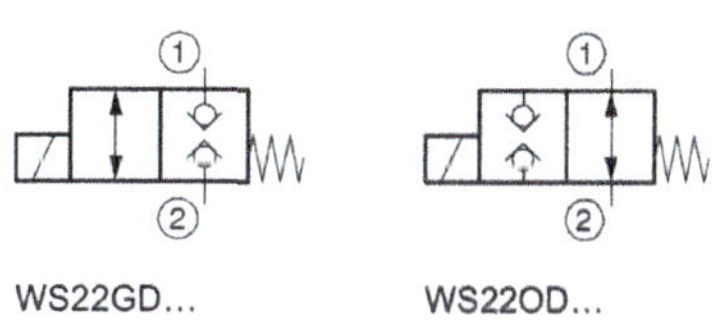

(a) (b)

Fig. 8.6 3D illustration of a Bucher Hydraulics on/off valve. WS22NG5Z (**a**). Diagram symbols from the valve data sheet. Normally closed and open are seen left and right respectively (**b**) [2]

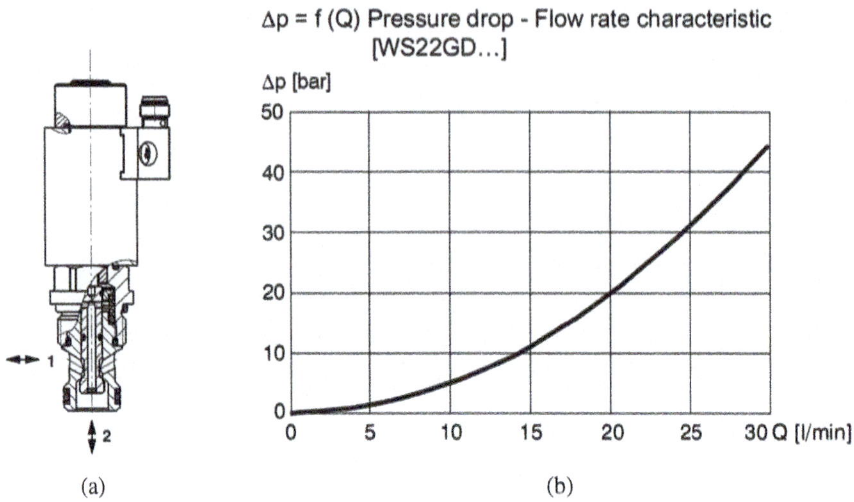

Fig. 8.7 Section cut of the poppet and seat part of the WS22 normally closed valve (**a**). Flow curve for the fully open WS22 normally closed valve (**b**) [2]

8.2.3 *Directional Spool Valve*

In spool valves, the moving part is a spool that moves in a valve hosing. By setting the spool in various positions, flow paths are created and controlled in size. In Fig. 8.8 a picture shows a spool valve that is cut open such that one see the spool and the internal connections. It can be seen how the spool is milled such that flow paths between the various ports are created as the spool is moved.

Let's examine the symbols in Fig. 8.9 to better understand the function of directional spool valves. Valves may have more that to port connections to fluid lines and they may control more than one flow path. The well known 4/3 way valve, for example, has 4 port connections and three positions—see top row far right in Fig. 8.9.

Fig. 8.8 Illustration of a spool valve

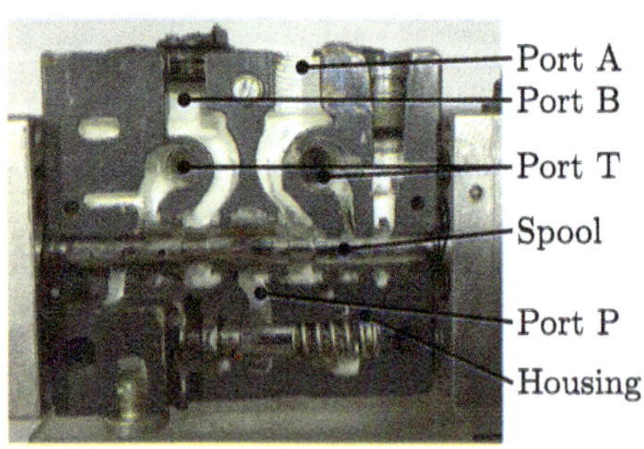

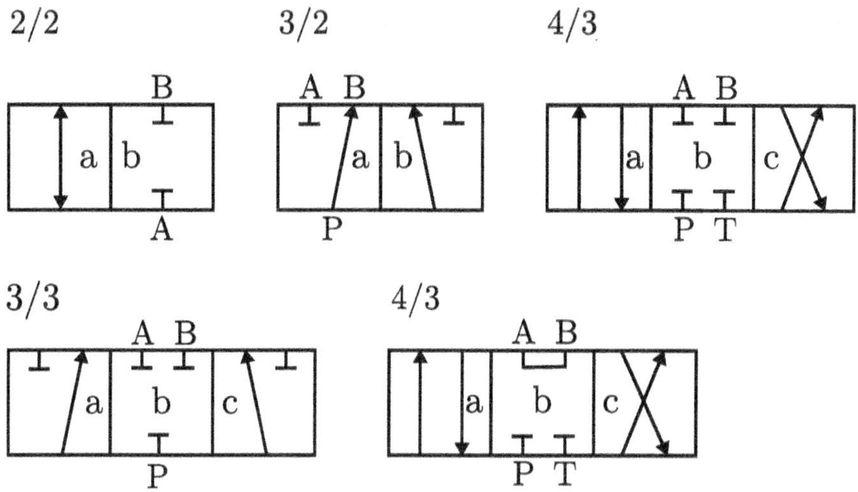

Fig. 8.9 Diagram illustration of directional valves

Depending on the chosen position—a, b or c—the valve connects port P to A and T to B, or P to B and T to A, for a and c respectively. Position c will close the flow paths.

For directional valves the "normal" or "neutral" position is the one with the connection annotation i.e. the 2/2 way on/off valve in Fig. 8.9 is normally open.

Directional valves may be constructed in almost all thinkable configurations such that the design engineer has multiple possibilities for creating flow paths between valve ports. In Fig. 8.10 some of the common valve states are depicted. A valve configuration is then a combination of a number of states. If the system designer requires the directional valve to restrict flow and not simply to direct it, this is indicated with an orifice symbol on the path.

Various centre or neutral states are shown in the bottom row where some of the connections are short circuited. We shall discuss some of these centre configurations later and see which to use when.

The valve state shifting is performed by various types of actuations depending on the use and size of the valve. Figure 8.11 shows the diagram symbols for four commonly used actuation types; manually, electromagnetic, pneumatic, fluid power.

Two 4/3 way valves are given in Fig. 8.12. They differ by the actuation but also by the two horizontal lines above and below the valve shown right. These lines indicate that the valve is a proportional valve. The proportional valve may be positioned partly into state a or c such that the opening area is proportional with the spool position. For a 4/3 way proportional valve as the one shown in Fig. 8.12a the flow area will be a function of the valve spool position, i.e., there is a proportionality between the spool position and the opening area.

In addition to the valve actuation, springs are often placed in the valve to manage the neutral position when no actuation force is applied. For both valves in Fig. 8.12

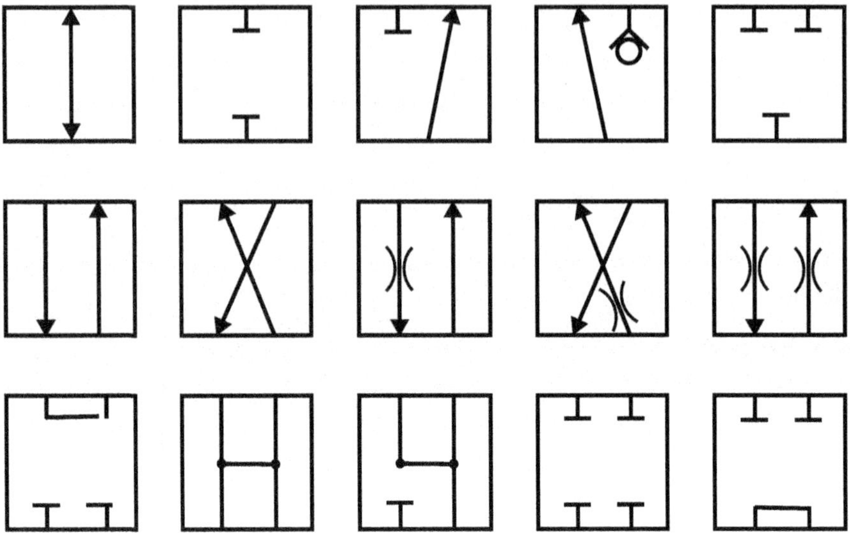

Fig. 8.10 Illustration of various direction valve spool configuration possibilities

Fig. 8.11 Valve actuation symbols

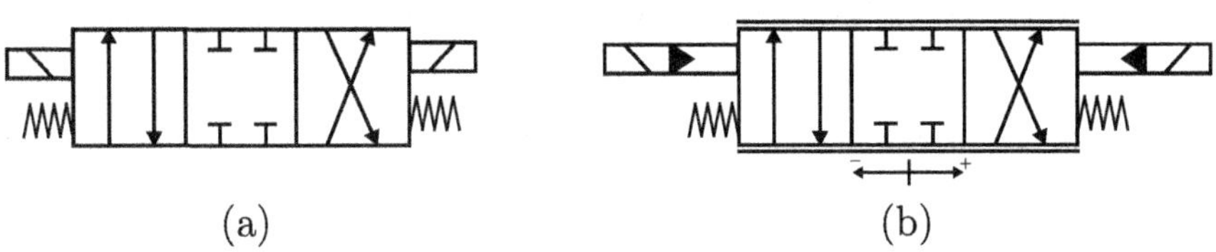

Fig. 8.12 Diagram symbols for a 4/3 way directional valve and a 4/3 way proportional valve, both spring centre

springs are placed to ensure that the valve is in the centre position when no actuation force is applied. The actuation for the directional valve is electrical. The proportional valve is electro-hydraulically actuated.

Valve Lap

Contrary to seat valves, spool valves can not be made leakage free since some clearance is required for the spool movement in the valve hosing. Leakage may, however, be lowered by using an overlap valve, see Fig. 8.13. For an overlap valve the flow path through the valve is closed for the first e.g. 10% spool movement, until the spool edge reaches the land edge.

Overlap valves, however, have a non-linear flow area function which complicates the controller design for systems using the proportional valve for actuator control.

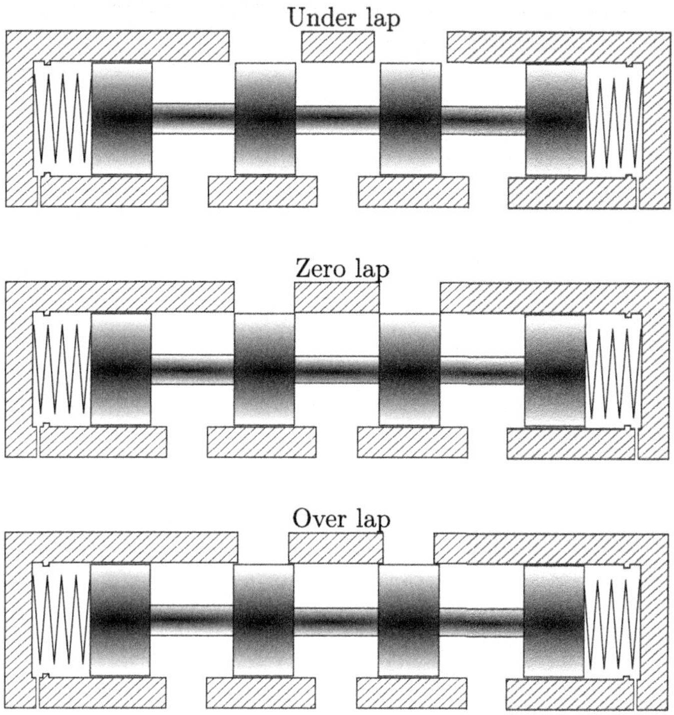

Fig. 8.13 Illustration of valve lap

Zero lap is when the spool edges and the land edges are aligned at neutral valve position. One may even ask for an under lap valve; the three valve configurations are seen in Fig. 8.13.

Which valve lap to choose is dependent on the system function. One may choose over lap to decrease the valve leakage or one may choose under lap for functionality reasons.

8.2.4 Flow Force on Spool Valve

In the Fluid Mechanics Part we saw how the momentum of fluids may be changed by external forces. In particular, in Sect. 3.2.3 we saw how changing the fluid flow direction changed the momentum which required external forces. When fluid flows through valves, flow directions are changed, leading to a requirement for an external force which is imposed by the valve housing or/together with the valve spool/poppet.

In this section we shall investigate the flow force induced in a spool valve with only one flow path. We investigate the simple spool valve in Fig. 8.14 where fluid is flowing from port P to A. The force requested to hold the valve spool is F_{R}.

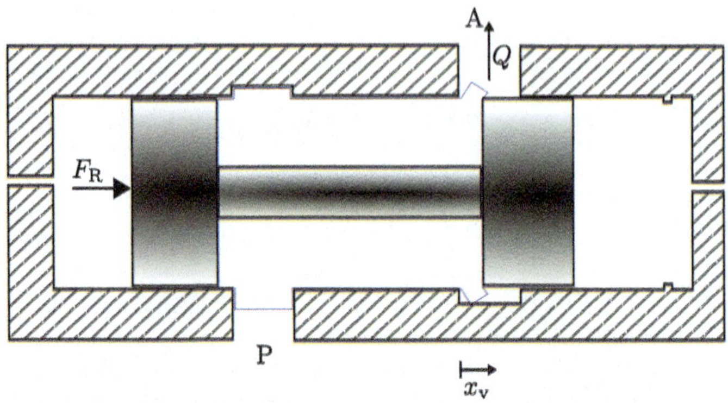

Fig. 8.14 Illustration of flow forces on spool valves

Fig. 8.15 Illustration of flow forces on spool valves

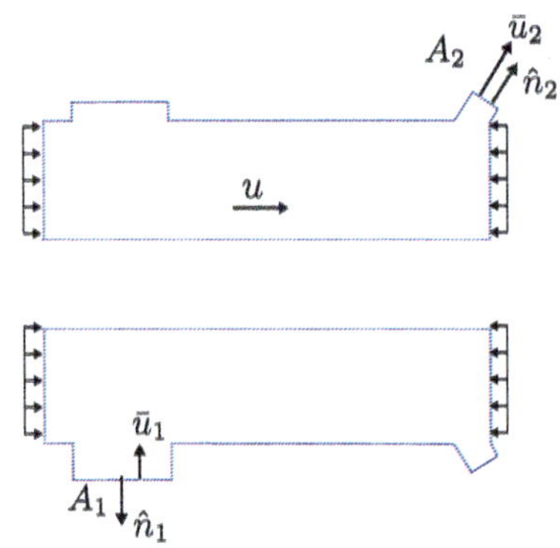

The control volume in Fig. 8.15 is defined such that the flow direction at the in and outlet is perpendicular to the flow area. Hence, the area normal vector and the fluid velocity are parallel at point 1 and 2.

As in Sect. 3.2.3 the momentum change is investigated using Reynolds Transport Theorem,

$$\Sigma F = \frac{D}{Dt}(mu), \tag{8.5}$$

$$F_B + F_S = \frac{\partial}{\partial t}\int_{cv} \rho u dV + \int_{cs} \rho u \bar{u} \cdot \hat{n}\, dA. \tag{8.6}$$

As the spool force is only applied in the x−direction the momentum equation for x−direction is of interest. We separate the right hand side in two terms, firstly momentum change inside the control volume and secondly the momentum crossing the surface.

Flow is crossing the surface at P and A ports. Note how the control volume surfaces are aligned perpendicular to the flow, such that the momentum crossing the surface may be written as:

$$\int_{\text{cs}} \rho u \bar{u} \cdot \hat{n}\, dA = \rho u_1 \bar{u}_1 \hat{n}_1 A_1 + \rho u_2 \bar{u}_2 \hat{n}_2 A_1, \tag{8.7}$$

$$= -\rho u_1^2 A_1 \cos\theta_1 + \rho u_2^2 A_2 \cos\theta_2. \tag{8.8}$$

The first term with the momentum change inside the control volume yields:

$$\frac{\partial}{\partial t}\int_{\text{cv}} \rho u dV = \frac{\partial}{\partial t}(\rho u L A_{\text{n}}) \tag{8.9}$$

The fluid velocity in the control volume is given by the flow and the flow area as: $u = \frac{Q}{A_{\text{n}}}$. Furthermore, the density is assumed constant, such that:

$$\frac{\partial}{\partial t}\int_{\text{cv}} \rho u dV = \rho L \frac{\partial}{\partial t}\left(\frac{Q}{A_{\text{n}}} A_{\text{n}}\right) \tag{8.10}$$

The flow through the valve may be modelled with the orifice equation:

$$Q = C_{\text{d}} w x_{\text{v}} \sqrt{\frac{2}{\rho}(p_{\text{P}} - p_{\text{A}})}, \tag{8.11}$$

where the opening area A_0 is given as the area gradient w multiplied by the spool position x_{v}, while the pressure drop is from the valve inlet to the valve outlet. The time derivative of the flow is given as:

$$\frac{\partial}{\partial t}(Q) = C_{\text{d}} w \sqrt{\frac{2}{\rho}(p_{\text{P}} - p_{\text{A}})} \frac{dx_{\text{v}}}{dt} + \frac{C_{\text{d}} w x_{\text{v}}}{\sqrt{2\rho(p_{\text{P}} - p_{\text{A}})}} \frac{d(p_{\text{P}} - p_{\text{A}})}{dt} \tag{8.12}$$

The total momentum change is given as (8.8) and (8.8) combined,

$$\frac{D}{Dt}(mu) =$$
$$\rho L C_{\text{d}} w \sqrt{\tfrac{2}{\rho}(p_1 - p_2)} \tfrac{dx_{\text{v}}}{dt} + \tfrac{\rho L C_{\text{d}} w x_{\text{v}}}{\sqrt{2\rho(p_1 - p_2)}} \tfrac{d(p_1 - p_2)}{dt} + \rho u_2^2 A_2 \cos\theta_2. \tag{8.13}$$

This momentum change must equal the sum of external body and surface forces, $F_{\text{B}} + F_{\text{S}}$. As the valve is placed horizontally the gravitational force is zero, which is why the only external force working on the control volume is the force from the spool on the fluid F_{R}. Note, however, that the force on the spool from the fluid is $-F_{\text{R}}$. Let's denote the force on the spool from the fluid as the flow force:

$$F_{\text{fl}} = -F_{\text{R}} \tag{8.14}$$

The flow force on the spool from the fluid flow through the valve is given as:

$$F_{\mathrm{fl}} = -\rho u_2^2 A_2 \cos\theta_2 \tag{8.15a}$$

$$-\rho L C_{\mathrm{d}} w \sqrt{\frac{2}{\rho}(p_1 - p_2)} \frac{dx_{\mathrm{v}}}{dt} \tag{8.15b}$$

$$-\frac{\rho L C_{\mathrm{d}} w x_{\mathrm{v}}}{\sqrt{2\rho(p_1 - p_2)}} \frac{d(p_1 - p_2)}{dt}. \tag{8.15c}$$

The flow force is given here in three parts; firstly the steady term (8.15a) which is depending on the size of the flow, fluid density and the outlet angle. The second term is the transient term with respect to change in valve spool position, i.e., the momentum change due to acceleration of the fluid volume given by LA_{n}. The third term is the transient term with respect to pressure change, hence as the pressure changes the amount of flow changes; this change calls for an acceleration of fluid.

Often the transient terms of the flow force are negligible and the steady term may be rewritten by assuming the discharge and the contraction coefficient to be equal, such that:

$$A_2 \approx C_{\mathrm{d}} A_0 \quad \text{and} \quad u_2 = \tfrac{Q}{A_2}, \tag{8.16}$$

$$F_{\mathrm{fl.ss}} = -\rho u_2^2 A_2 \cos\theta_2 = -\rho Q \tfrac{Q}{A_2} \cos\theta_2, \tag{8.17}$$

$$= -\rho C_{\mathrm{d}} A_0 \sqrt{\tfrac{2}{\rho}(p_1 - p_2)} \frac{A_2 \sqrt{\frac{2}{\rho}(p_1 - p_2)}}{A_2} \cos\theta_2, \tag{8.18}$$

$$F_{\mathrm{fl.ss}} = -2 C_{\mathrm{d}} A_0 (p_1 - p_2) \cos\theta_2. \tag{8.19}$$

The flow through this spool valve applies a force in the spool seeking to close the valve even for a steady flow.

Note that the flow force computed here is for **this specific control volume**, and for another valve another flow force equation may appear.

8.2.5 Servo Valves

Servo valves are merely proportional valves, however, the word servo implies that the valve is controlled. In servo valves the valve spool position may be controlled in various ways. Basically, the position of the main spool is electrically or mechanically measured and compared to a reference value given as input; the difference between required position and actual position is utilised as controller input. The control then

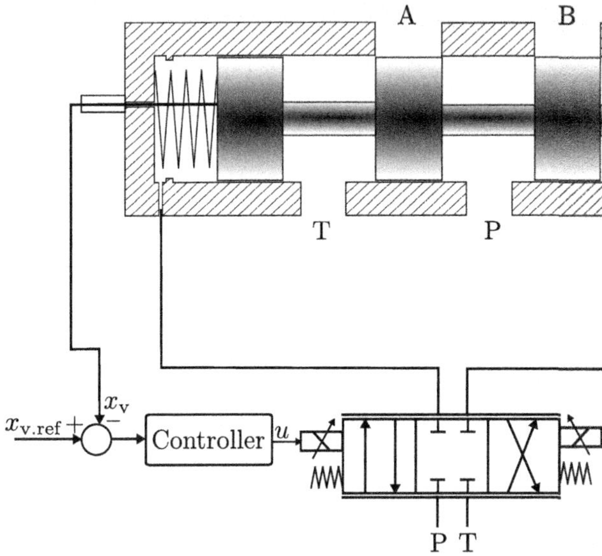

Fig. 8.16 Illustration of flow forces on spool valves

calculates the requested actuation force to be implemented. The actuation force may be applied in various ways, e.g. electrically with solenoid or voice coil actuators, or with fluid pressure controlled by another valve setup.

A servo valve in which the main spool position is controlled with fluid pressure via another valve is often named a two stage valve. The first stage is controlling the spool position of the second stage, as the position in a cylinder is controlled. A two stage servo valve is seen illustrated in Fig. 8.16 where the pilot valve (first stage) is used to control the position of the main spool. In the depicted configuration the main spool position is measured with a position sensor and the signal is compared to the requested spool position. The position error is used in the controller calculating the pilot spool position. Note that the pilot spool controls flow into and out of the two pilot chambers at both ends of the main spool; in this way the pilot spool controls the main spool position.

The first or pilot stage of the two stage valve in Fig. 8.16 is a 4/3-way proportional valve, actuated with a electrical actuator capable of delivering a variable (indicated by the arrow) force. The control loop, in terms of position measurement, reference signal and valve signal to the pilot spool, is as such a fully electrical loop.

In the following, two industrial servo valves are introduced to illustrate how the valves function and construction may be accomplished. The chosen valves are used in laboratory experiments at AAU ENERGY and the first one is utilised in some of the laboratory exercises accompanying the course given at AAU ENERGY.

Example of Industry Valve—Moog D633

The Moog D633 servo valve is used to illustrate a direct drive servo valve with position feedback. A cut-through section view is given in Fig. 8.17. It is a block mount type valve such as is designed to be mounted on a valve block which has the four(five) connections T, A, P, B and Y as seen on the figure. The Y port is important in cases with high pressure on the tank port. The T connection is internally connected to both ends of the valve's spool, such that A and T are connected when the spool is moved right and B and T are connected when the spool is moved left. The P connection is placed in the middle such that it also connects to either A or B opposite the T connection.

In Fig. 8.17 3 is the spool, 6 is the Linear Force Motor and 5 is the centering spring. 7 and 8 indicate the position sensor and integrated electronics used in controlling the spool position. With the integrated electronics and force motor onboard the valve unit, the Valve connector 2 includes power for actuation, a reference signal and a position measurement signal.

The valve is a 4/3-way proportional valve as seen in the diagram symbol in Fig. 8.18. The diagram symbol further indicates the electronic position feedback and control with the "lightning" arrow.

For modelling purposes the most interesting parts of the data sheet are the curves for valve flow and spool dynamics; these are seen in Figs. 8.18a and 8.19b respectively.

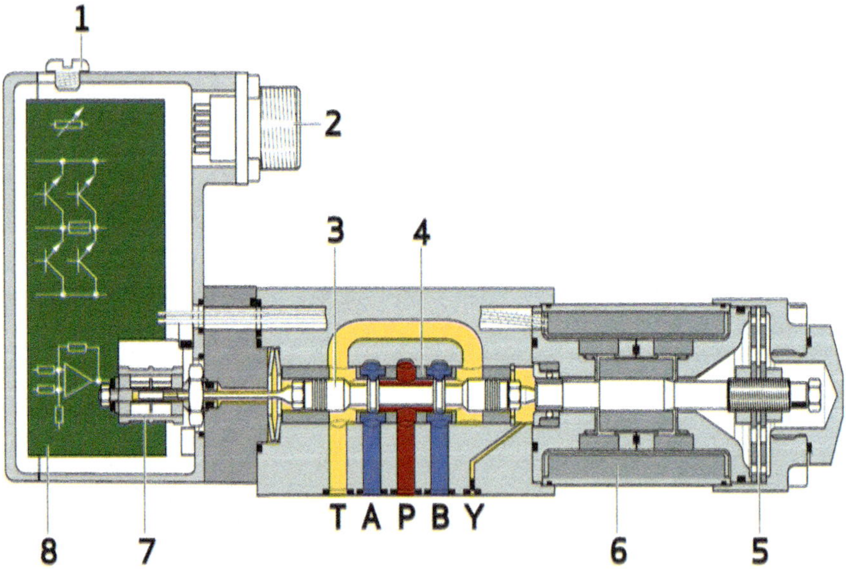

Fig. 8.17 Section view of the Moog D633, from the data sheet ®Moog Inc [3]

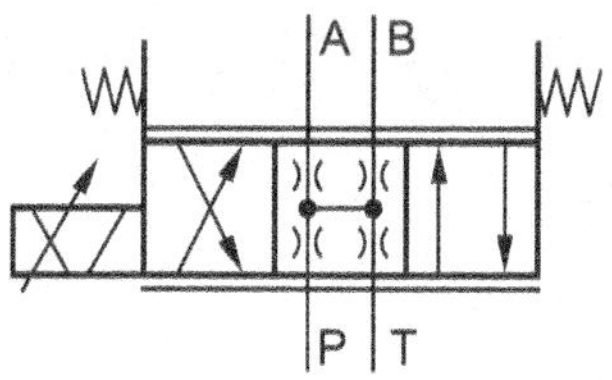

Fig. 8.18 Diagram symbol for the Moog D633, from the data sheet ®Moog Inc [3]

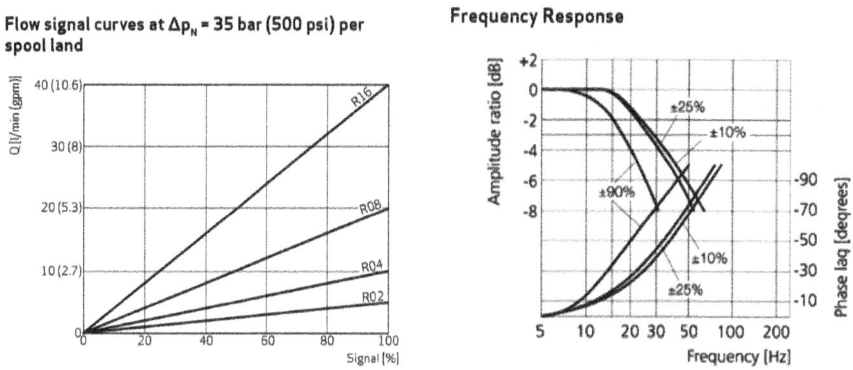

(a) (b)

Fig. 8.19 Flow curves and frequency response for the Moog 633 valve as given in the data sheet®Moog Inc [3]

In the data sheet, Eq. (8.20) is given for the flow through a fully open valve. The flow Q through the valve at a pressure drop Δp is such that,

$$Q = Q_{\mathrm{N}} \sqrt{\frac{\Delta p}{\Delta p_{\mathrm{N}}}}, \tag{8.20}$$

where Q_{N} and Δp_{N} are the nominal flow and pressure for the fully open valve respectively. Combining Eq. (8.20) and the curve in Fig. 8.19a one may set up the valve flow as,

$$Q(t) = \frac{Q_{\mathrm{Nom}}}{\sqrt{\Delta p_{\mathrm{Nom}}}} \frac{x_{\mathrm{v}}(t)}{x_{\mathrm{v,max}}} \sqrt{\Delta p(t)}, \tag{8.21}$$

where $x_v(t)$ is the spool movement such that $\frac{x_v(t)}{x_{v,max}}$ is the nominal valve opening with 1 corresponding to a valve signal of 100%. The flow in Eq. (8.20) is such that the flow from e.g. P to A or B to T etc. depending on if the spool is moved left or right in the housing.

From the frequency response in Fig. 8.19b one may get the idea of modelling the valve movement with a second order system as,

$$\frac{X_v(s)}{U(s)} = \frac{\omega_{n.v}^2}{s^2 + 2\zeta_{n.v}\omega_{n.v}s + \omega_{n.v}^2}, \tag{8.22}$$

where $U(s)$ is the control signal, $\omega_{n.v}$ and $\zeta_{n.v}$ are the natural frequency and damping for the valve movement respectively. It is worth noting that the bandwidth of the valve is dependent on the amplitude of control signal as naturally expected for a non-linear system with actuation limits.

In later chapters we shall develop a system in which the valve model plays an important role; we shall discuss if modelling the valve dynamics is important or a waste of resources.

Example of Industry Valve—Moog D761

The Moog 671 is used to illustrate a flapper nozzle type servo valve. Flapper Nozzle servo valves feature mechanical feedback from the spool position to the spool actuation. The spool actuation is by means of fluid power, such that the main spool is moved by fluid pressure. A cutaway illustration of the Moog 631 is seen in Fig. 8.20.

The pilot valve controlling the pressure in the two pilot chambers is composed of two nozzles, each fed with fluid through one of the pilot chambers situated at each end of the spool. In other words, if the flow through one of the nozzles is high, the pressure in the connected pilot chamber is low. Each pilot chamber is fed from the P (or X) port through a fixed orifice such that this and the nozzle are two restrictions in series. With the nozzle restriction being controlled by the flapper position, pressure drop is distributed between the two restrictions while controlling the pilot chamber pressure. The pilot force is in this way controlled by the flow through the two nozzles. The flapper is moved by the torque motor on top and the spool movement on the bottom.

By moving the flapper closer to a nozzle, the flow through that nozzle decreases while increased in the one opposite. Low nozzle flow yields high pilot pressure, imposing force on the spool. The spool movement drags the flapper away from the nozzle and an equilibrium between applied torque and spool position is established. A given torque on the flapper corresponds to a given spool position.

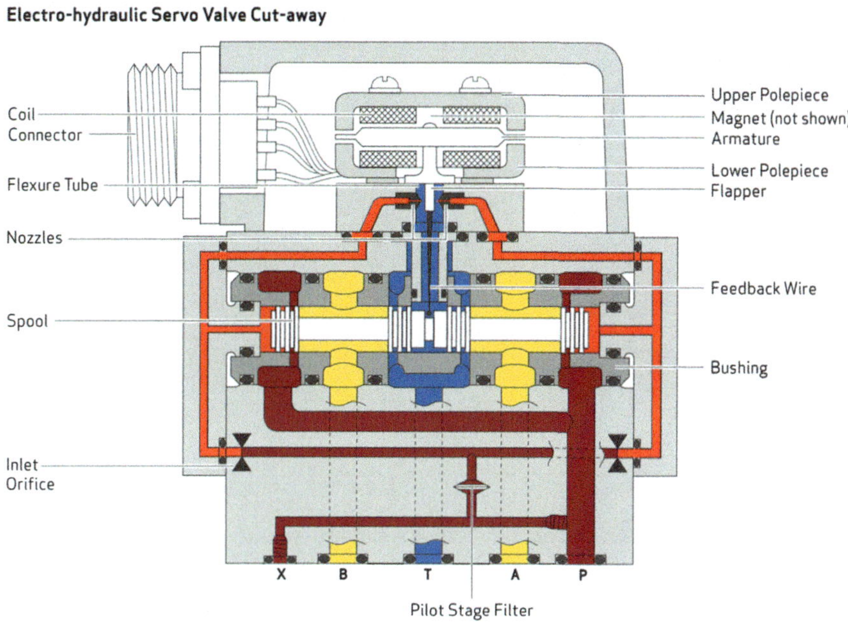

Fig. 8.20 Cutaway illustration of the Moog D761 flapper nozzle valve, ®Moog Inc [4]

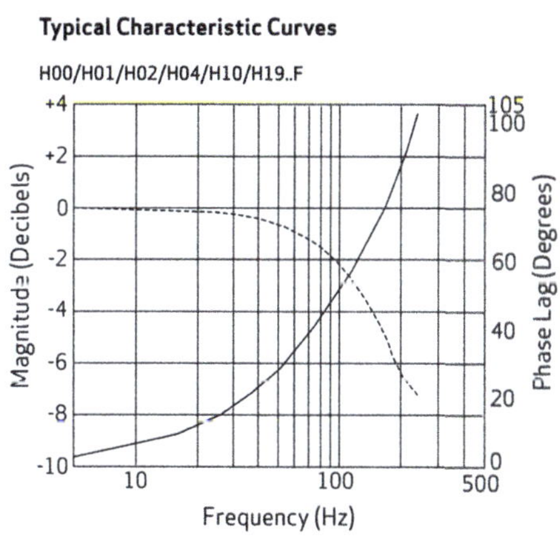

Fig. 8.21 Frequency response for the Moog D761 flapper nozzle valve, ®Moog Inc [4]

The flapper nozzle design enables fast valves, however, with a limited stroke length. The frequency response of the Moog D761 flapper nozzle valve is seen in Fig. 8.21. The current valve has a bandwidth of approximately 100 Hz; in the data sheet for this valve series both slower and faster responses are seen.

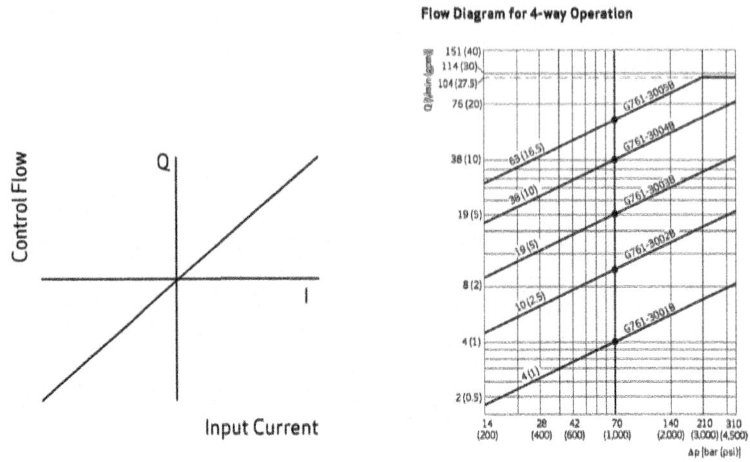

(a) (b)

Fig. 8.22 Flow curves for the Moog D761 valves. **a** illustrates the input-control flow relation and **b** shows the pressure flow relation for a fully open valve, ©Moog Inc [3]

The data sheet for the Moog D761 states the same flow equation as for the Moog D633:

$$Q = Q_N \sqrt{\frac{\Delta p}{\Delta p_N}}, \tag{8.23}$$

where Q_N and Δp_N are the nominal flow and pressure for the fully open valve respectively. This is supported by the curve in Fig. 8.22b and with the input current to control flow relation given as seen in Fig. 8.22a one arrives at a flow equation given as:

$$Q(t) = \frac{Q_{Nom}}{\sqrt{\Delta p_{Nom}}} \bar{I}(t) \sqrt{\Delta p(t)}, \tag{8.24}$$

where $\bar{I}(t)$ is the normalised input current.

8.3 Pressure Control Valves

Pressure control valves control the fluid pressure in a given pressure node (control volume); this node can be upstream or downstream of the pressure control valve. In this lecture note, three pressure function valves are introduced: namely, the pressure relief, reduction and control valves.

8.3.1 *Pressure Relief*

Pressure protection valve is perhaps a more descriptive name for the pressure relief valve, as its main purpose is to protect the fluid power system against destructive high pressure.

"All" fluid power systems should be equipped with a pressure relief valve just downstream of the pump, such that if e.g. a fixed displacement constant velocity pump is displacing fluid towards a closed directional valve, the relief valve will eventually open and direct fluid back to the tank before the pressure becomes hazardous.

The single most important value for a pressure relief valve is the cracking pressure, which is the pressure level at which the valve opens. Other important parameters are the flow capacity and in some cases eigenfrequency of the valve. Relief valves are constructed in several ways; here we shall present a one stage relief valve.

One Stage Relief Valve

Diagram symbols for two pressure relief valves are seen in Fig. 8.23. Relief valves are normally closed valves, which is indicated as the flow arrow is not aligned with the port connections A and B. The B port is often connected to the system tank. A hydro-mechanical relief valve features a spring applying force to keep the valve closed; on the other hand, pilot pressure from e.g. A applies force on the opposing site trying to open the valve. Various pilot configurations may be constructed, the two basic ones, though, are seen in Fig. 8.23.

Valve (a) and (b) in the figure differ by the latter having an adjustable spring pre-compression and as such, an adjustable cracking pressure. Furthermore, the pilot area accompanying the spring force is connected in the (b) valve to the B port, such that the valve opens at a given relative pressure difference between the A and B pressure. The (a) valve conversely opens at a given A pressure relative to gauge pressure, as the pilot accompanying the spring is vented to ambient.

Simple illustrations of how the two discussed valve symbols may be constructed are given in Fig. 8.24. The pre.compression of the spring is handled with a screw such that one may manually adjust the compression of the spring and hence the force applied on the poppet when the valve is closed.

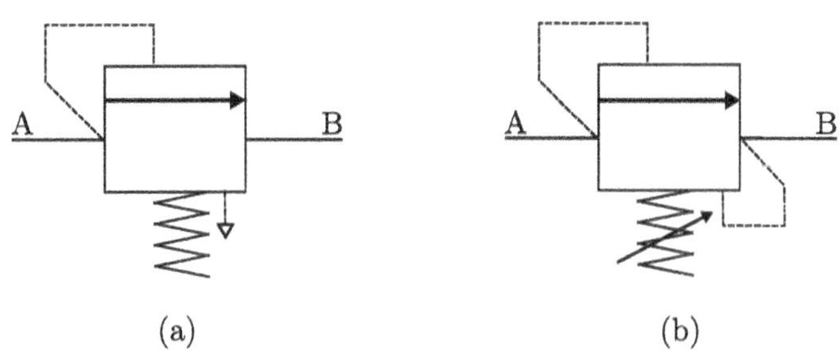

Fig. 8.23 Illustration of two single stage pressure relief valves with fixed and variable cracking pressure respectively

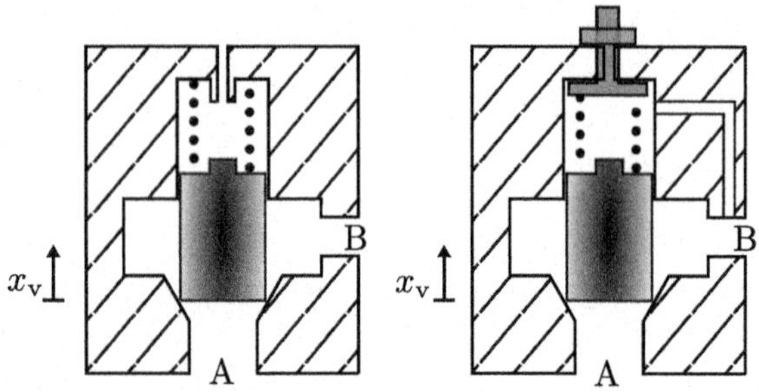

Fig. 8.24 Section cut illustrations of two single stage pressure relief valves with fixed and variable cracking pressure respectively

Let's use the general valve model setup from (8.1) and (8.2) to setup models for the two pressure relief valves (a) and (b) in Fig. 8.24, first (a) where the flow through the valve is

$$Q = C_d A_0(x_v)\sqrt{\frac{2}{\rho}|p_A - p_B|}\frac{p_A - p_B}{|p_A - p_B|}, \tag{8.25}$$

where C_d is the valve discharge coefficient, rho is the oil density, p_A and p_B are the pressure at port A and B respectively and $A_0(x_v)$ is the flow area function depending on the valve poppet position x_v. We model the poppet movement with Newton's Second Law as:

$$\ddot{x}_v M_v = A_A p_A - F_{fr}(\dot{x}_v) - k_s(x_{v.0} + x_v) + F_{seat} + F_{fl}, \tag{8.26}$$

where M_v is the poppet mass, A_A is the pilot area on which the A port pressure works, k_s is the spring constant, which together with the pre-compression $x_{v.0}$, set the spring force on the closed valve. Due to the mechanical contact between seat and poppet a seat force, F_{seat}, is present when the valve is closed. The flow force, F_{fl}, is on the other hand only present when fluid is flowing through the valve.

We see that the flow direction is defined as positive from A to B. Furthermore, it is evident that the forces working on the poppet do not include B side pressure or pilot pressure forces, other than from port A, as well as no actuation force being applied. A seat force is here applied as one must have some kind of end stop model for the poppet; if not included, the poppet equilibrium point will be well down into port A. This may, however, be solved with a numerical limit in a simulation model such that a contact force model may be omitted for the seat force.

For the (b) valve we need to include the B port pressure in the momentum equation while the orifice equation for the flow stays the same:

$$Q = C_\mathrm{d} A_0(x_\mathrm{v}) \sqrt{\frac{2}{\rho}|p_A - p_B|} \frac{p_A - p_B}{|p_A - p_B|}. \tag{8.27}$$

The momentum equation for the (b) valves is:

$$\ddot{x}_\mathrm{v} M_\mathrm{v} = A_A p_A - A_B p_B - F_\mathrm{fr}(\dot{x}_\mathrm{v}) - k_\mathrm{s}(x_\mathrm{v.0} + x_\mathrm{v}) + F_\mathrm{seat} + F_\mathrm{fl}, \tag{8.28}$$

where A_B is the pilot area on which the B port pressure works. The model of the two valves differs only by the inclusion of the B pressure force. However, note that the spring pre-compression $x_\mathrm{v.0}$ may be adjusted for valve (b) such that the A pressure level leading to initial valve movement is changed.

The cracking pressure for valves (a) and (b) respectively is given as:

$$p_\mathrm{cr} = \frac{k_\mathrm{s} x_\mathrm{v.0}}{A_A}, \tag{8.29}$$

$$p_\mathrm{cr} = \frac{k_\mathrm{s} x_\mathrm{v.0}}{A_A} + \frac{p_B A_B}{A_A}. \tag{8.30}$$

The cracking pressure equation clearly states that valve design (a) controls to absolute pressure and valve design (b) controls a relative pressure between A and B.

Example of Industry Valve—Parker A04B2

Pressure relief valves may be constructed in numerous ways, however in Fig. 8.25a a cartridge type pressure relief valve is depicted as given in the data sheet [5]. Note that the back pressure is included in the force balance such that it controls relative pressure. A screw is used to compress the spring.

From the curves in Fig. 8.25b one should note that no pressure relief valve is ideal why the pressure drop depends on the flow through the valve.

8.3.2 Pressure Reduction

The downstream pressure may be reduced with a pressure reduction valve. Hence, with a pump pressure or another higher upstream pressure the pressure reduction valve reduces to a pre-set pressure value just downstream of the valve. The diagram symbol of the pressure reduction valve is seen in Fig. 8.26 one notes that this valve is normally opened, indicated with the flow arrow aligned with the port connections, P and A (contrary to the pressure relief valve). A spring is placed to keep the valve opened while a pilot chamber connected to port A generates force working to close

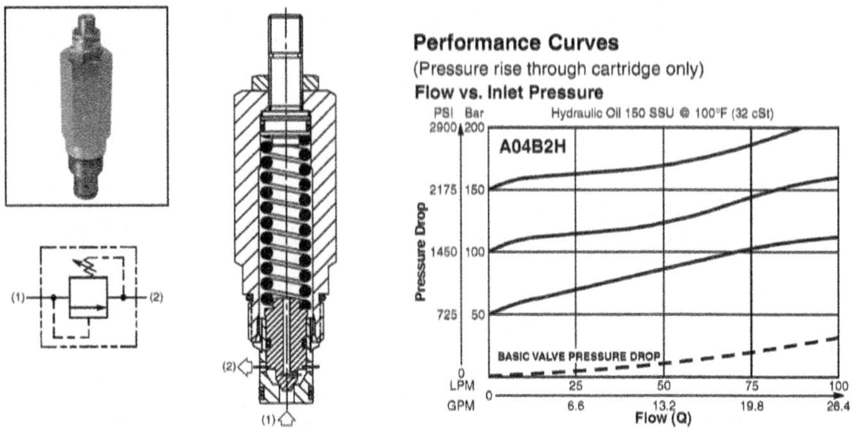

(a) (b)

Fig. 8.25 Illustration of the A04B2 pressure relief valve, picture, symbol, section view and flow curve, ®Parker Inc [3]

Fig. 8.26 Diagram symbol of a pressure reduction valve

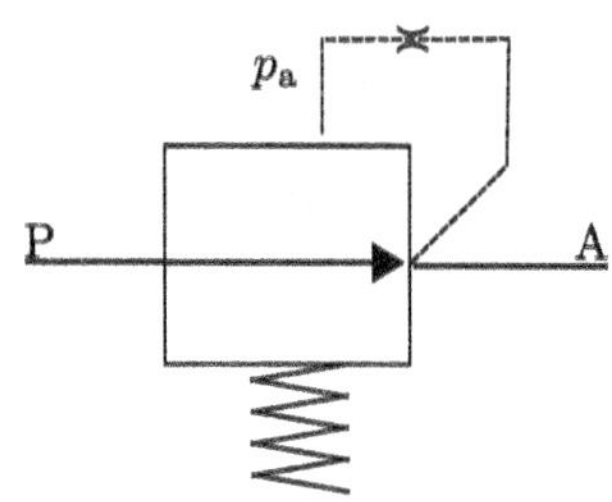

the valve. The balance between the spring and pressure force by the pressure p_a gives the position of the valve spool or poppet.

A section view of a simple pressure reduction valve is given in Fig. 8.27. The spring chamber is for now vented to tank pressure through a "large" opening, while the pilot chamber a is connected to port A through an orifice in which the flow Q_a runs. The main flow from port P to port A is named Q_{PA}.

Let's set up the describing equations for the pressure reduction valve in Fig. 8.27. Firstly, the orifice equation for the main flow,

$$Q_{PA} = C_d A_{PA}(x_v)\sqrt{\frac{2}{\rho}|p_P - p_A|}\frac{p_P - p_A}{|p_P - p_A|}, \tag{8.31}$$

where $A_{PA}(x_v)$ is the flow area depending on spool position x_v and p_P and p_A are the pressures at port P and A respectively. Note, that the flow area function $A_{PA}(x_v)$ is zero for negative spool positions and linear proportional to the spool position for the circular spool shown. C_d and ρ are once again valve discharge coefficient and oil density. Secondly, the momentum equation for the spool is,

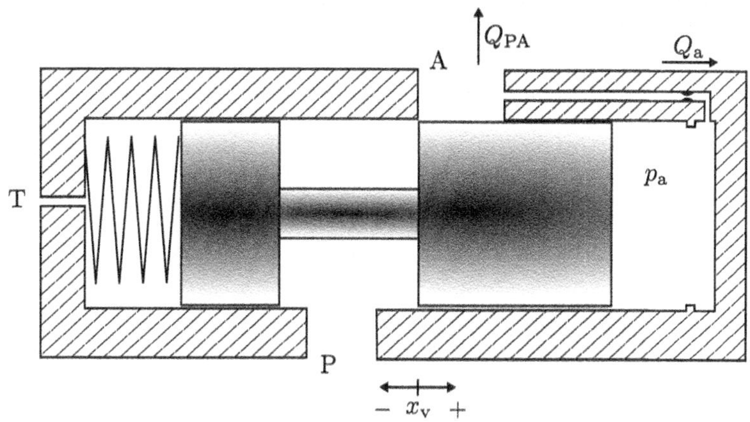

Fig. 8.27 Section view of a simple pressure reduction valve

$$\ddot{x}_v M_v = A_v p_a - A_v p_T - F_{fr}(\dot{x}_v) + k_s(x_{v.0} - x_v) + F_{fl}, \tag{8.32}$$

where the P port pressure does not influence the spool movement as it works on the area A_P in both positive and negative direction. Note that A_v is end area of the spool, which, if the spool diameter is d, is given as $A_v = d^2\frac{\pi}{4}$. k_s is the spring constant which, together with the pre-compression $x_{v.0}$, set the spring force on the spool when in neutral position ($x_v = 0$).

The pilot pressure, p_a, is differing from the port pressure, p_A, due to the small damping orifice leading to the following two equations related to the pilot chamber,

$$Q_a = C_d A_a \sqrt{\frac{2}{\rho}|p_A - p_a|}\frac{p_A - p_a}{|p_A - p_a|}, \tag{8.33}$$

$$\dot{p}_a = \frac{\beta(p_a)}{V_{a0} - x_v A_v}\left(Q_a + A_v \dot{x}_v\right), \tag{8.34}$$

where V_{a0} is the volume of the pilot chamber with the valve spool in zero position. $\beta(p_a)$ is the effective bulk modulus of the fluid in the pilot chamber and A_a is the flow area in the orifice connecting port A and the pilot chamber.

8.3.3 Pressure Control

Pressure control valves are designed to control the pressure to a given pressure level at a point in the fluid power system, e.g. in a cylinder chamber. Contrary to the pressure reduction valve the pressure control valve may lead flow away from the controlled point, which is why the pressure control valve has three main ports. A diagram symbol for a pressure control valve is seen in Fig. 8.28.

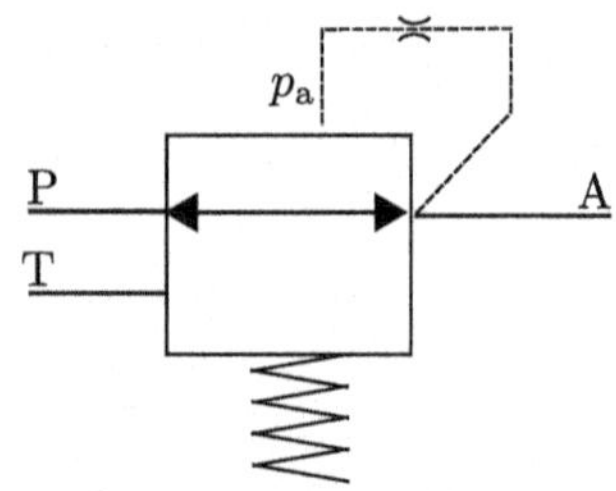

Fig. 8.28 Diagram symbol of a pressure control valve

From the symbol it can be seen that the pressure control valve is normally open until the required pressure is reached. As the controlled pressure gets closer to the requested pressure the valve spool moves toward $x_v = 0$. At the point where the A pressure is as requested, the spring pre-compression force and the pilot pressure force are equal, and $x_v = 0$. This is, however, only possible if there is no flow through port A, as a flow through the valve requires the valve spool to be slightly shifted to one side. In other words, as long as a flow is flowing though port A the A pressure is not equal to the requested pressure, but it can be close to it for a properly designed pressure control valve.

Figure 8.29 illustrates how a pressure control valve according to the given symbol may be constructed as a spool valve. An orifice is often placed between the controlled pressure (port A) and the small pilot chamber in the pressure control valve to improve damping of the valve. How this effects the damping and eigenfrequency of the valve spool, we shall discuss in a later chapter under system and component analysis. Here, we note that the pressure in the pilot chamber differs from the A pressure when a flow through the small damper orifice is present.

Let's set up the describing equations for the pressure control valve in Fig. 8.29. Again we start with the orifice equation for the main flow,

$$Q_A = C_d A_{PA}(x_v)\sqrt{\frac{2}{\rho}|p_P - p_A|}\frac{p_P - p_A}{|p_P - p_A|} + C_d A_{TA}(x_v)\sqrt{\frac{2}{\rho}|p_T - p_A|}\frac{p_T - p_A}{|p_T - p_A|} \tag{8.35}$$

where the area functions $A_{PA}(x_v)$ and $A_{TA}(x_v)$ both depend on the spool position x_v. Furthermore, they are zero for negative and positive spool positions respectively. The momentum equation for the spool is,

$$\ddot{x}_v M_v = A_v p_a - A_v p_T - F_{fr}(\dot{x}_v) + k_s(x_{v.0} - x_v) + F_{fl}, \tag{8.36}$$

where M_v is spool mass, k_s is spring constant and $x_{v.0}$ is spring pre-compression. A_v is the spool end area on which pilot pressures p_a and p_T respectively work. In addition to pressure and spring forces a velocity dependent friction force, $F_{fr}(\dot{x}_v)$, is included as well as a flow force, F_{fl}.

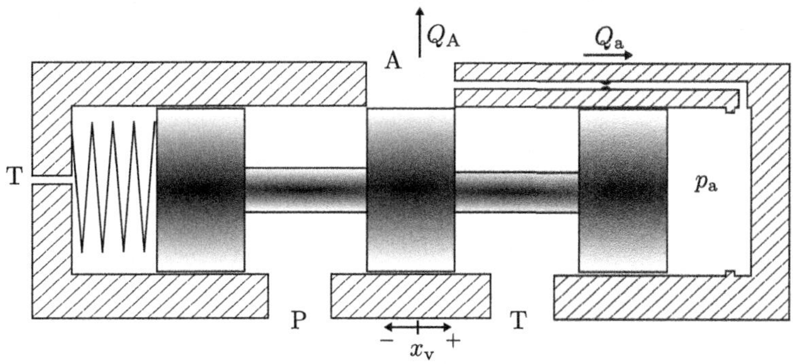

Fig. 8.29 Section view of a simple pressure control valve

The pressure control valve as it is designed in Fig. 8.29 requires a model for the pilot pressure and the flow into the pilot chamber given as,

$$Q_a = C_d A_a \sqrt{\tfrac{2}{\rho} |p_A - p_a|} \tfrac{p_A - p_a}{|p_A - p_a|}, \quad (8.37)$$

$$\dot{p}_a = \tfrac{\beta(p_a)}{V_{v.0} - x_v A_v} \left(Q_a + A_a \dot{x}_v\right), \quad (8.38)$$

where the orifice area A_a connecting port A and the pilot chamber, limits flow Q_a to tune the dynamics of the valve. V_{a0} is the volume of the pilot chamber with the valve spool in zero position. $\beta(p_a)$ is the effective bulk modulus of the fluid in the pilot chamber.

8.4 Flow Control Valves

The ability to control fluid flow is requested when one wants to control the velocity of fluid power actuators. We shall examine various means to flow control valves. Diagram symbols of valves utilised for flow control are seen in Fig. 8.30.

The first two are throttle valves represents "simple" restrictions in a flow path. Conversely the later two are throttle valves combined with a pressure function valve.

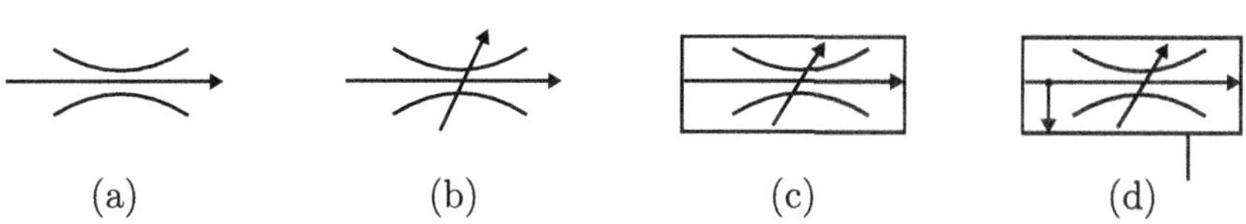

Fig. 8.30 Diagram symbol for flow controlling valves. **a** Fixed throttle valve. **b** Adjustable throttle valve. **c** Pressure compensated flow control valve. **d** Pressure compensated flow control valve—by pass

In the next four subsections we shall familiarise ourselves with fixed and adjustable throttle valve as well two pressure compensated flow control valves. Further to this, an example is given of how the two valves' categories may be used for piston velocity control.

8.4.1 Throttle Valve

Throttle valves are in some simple systems used to control the amount of fluid flow, i.e. they are used as flow control valves often to control actuator velocity. Normally the flow through throttle valves is turbulent, hence the orifice equation applies,

$$Q = C_\text{d} A_\text{d} \sqrt{\frac{2}{\rho}|\Delta p|} \text{sgn} \Delta p. \tag{8.39}$$

A fixed throttle valve, Fig. 8.30a, is a valve with a constant opening area, where the flow is depending solely on the pressure difference across the valve. An adjustable throttle valve, Fig. 8.30b, is a valve with an adjustable opening area, for which the opening area may be chosen in a range often from zero to $A_\text{d.max}$. With the adjustable throttle valve the area may be adjusted to the required flow amount for the current operation or system. However, during operation the flow through the throttle valves is still dependent on the pressure difference across the valve. Figure 8.31a shows a cartridge needle valve as depicted in the data sheet [6] (adjustable throttle valve). The current valve is adjustable with the hand screw at the top. A diagram showing the flow dependent on pressure drop and turns on the hand knob is given in Fig. 8.31b.

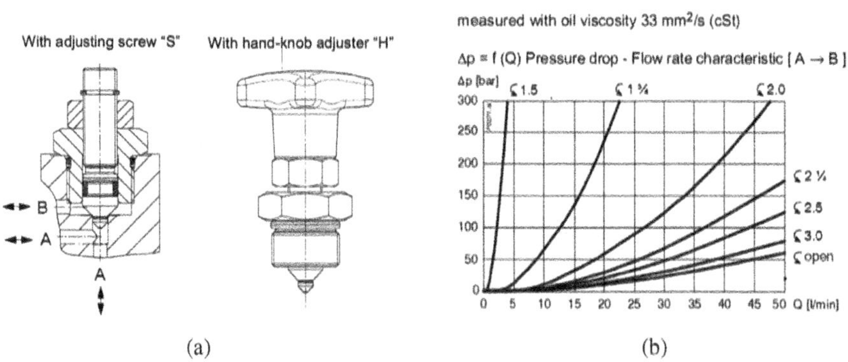

Fig. 8.31 Illustration of the Bucher MDPWA5S throttle valve, section view, drawing and flow curve for various screw positions, ® Bucher Hydraulics Inc [6]

8.4.2 Case Illustration—Throttle Valves

Let's examine a case exemplifying how throttle valves may be used for velocity control of a fluid power cylinder. In Fig. 8.32 a differential cylinder is seen in two cases; extending and retracting (left and right). The velocity is controlled in both directions by metering in control, i.e. the inflow is controlled while the outflow is not.

Left in Fig. 8.32 fluid is lead from supply (p_S) to the A side chamber through a fixed throttle valve, while the outflow from chamber B is flowing through the check valve. The 4/2 way directional valve is dictating the flow direction, however it is designed with close to no flow restriction, such that only a minimum pressure drop is experienced across the directional valve. Likewise, the pressure drop across the open check valve is close to zero. Right in Fig. 8.32 fluid is led from the supply (p_S) to the B side chamber through the adjustable throttle valve, while the outflow from chamber A is flowing through the check valve.

Let's now perform a steady state "analysis" of the system in Fig. 8.32 assuming the pressure drop across the open check valve and the 4/2 directional valve to be zero.

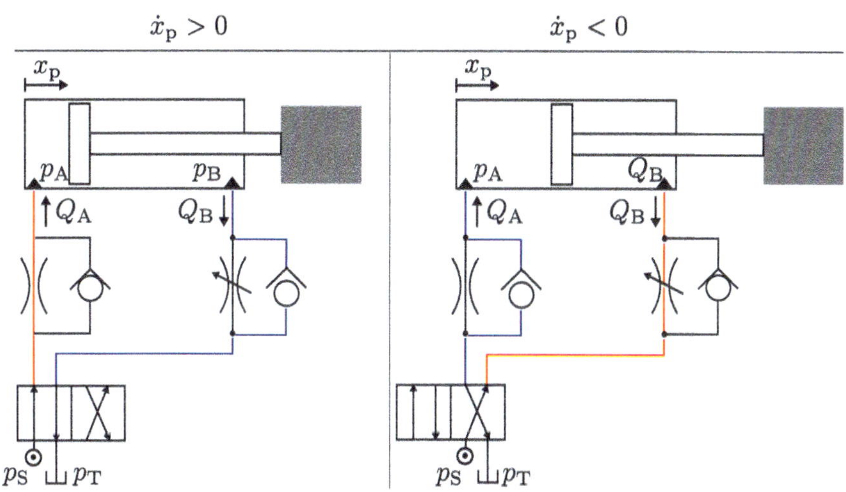

Fig. 8.32 Illustration of velocity control with throttle valves

8.4.2.1 Extending Piston

Fluid flow through the A side orifice is in the left hand side situation:

$$Q_{\mathrm{A}} = C_{\mathrm{d}} A_{\mathrm{d.A}} \sqrt{\frac{2}{\rho} |p_{\mathrm{S}} - p_{\mathrm{A}}|} \mathrm{sgn}(p_{\mathrm{S}} - p_{\mathrm{A}}), \tag{8.40}$$

where C_{d} and $A_{\mathrm{d.A}}$ are the valve's discharge coefficient and flow area, while $p_{\mathrm{S}} - p_{\mathrm{A}}$ is the pressure drop across the fixed orifice. The steady state continuity equation for chamber A is:

$$Q_{\mathrm{A}} = \dot{x}_{\mathrm{p}} A_{\mathrm{A}}, \tag{8.41}$$

with A_{A} being the piston area, the piston velocity is given as:

$$\dot{x}_{\mathrm{p}} = \frac{Q_{\mathrm{A}}}{A_{\mathrm{A}}}. \tag{8.42}$$

Combining this with the orifice equation the velocity may be written as:

$$\dot{x}_{\mathrm{p}} = \frac{C_{\mathrm{d}} \sqrt{\frac{2}{\rho}}}{A_{\mathrm{A}}} A_{\mathrm{d.A}} \sqrt{|p_{\mathrm{S}} - p_{\mathrm{A}}|}, \tag{8.43}$$

if the pressure drop is positive ($p_{\mathrm{S}} > p_{\mathrm{A}}$). The flow out of the piston on the B side is:

$$Q_{\mathrm{B}} = \dot{x}_{\mathrm{p}} A_{\mathrm{B}}. \tag{8.44}$$

As the A side throttle valve has a fixed opening area the velocity will change if the machine was used in a different loading condition.

Retracting Piston

Fluid flow through the B side orifice is in the right hand side situation:

$$Q_{\mathrm{B}} = C_{\mathrm{d}} A_{\mathrm{d.B}} \sqrt{\frac{2}{\rho} |p_{\mathrm{B}} - p_{\mathrm{S}}|} \mathrm{sgn}(p_{\mathrm{B}} - p_{\mathrm{S}}) \tag{8.45}$$

where C_{d} and $A_{\mathrm{d.B}}$ are the valve's discharge coefficient and flow area, while $p_{\mathrm{B}} - p_{\mathrm{S}}$ is the pressure drop across the fixed orifice. The steady state continuity equation for chamber B is setup and used to find the piston velocity:

$$-Q_{\mathrm{B}} = -\dot{x}_{\mathrm{p}} A_{\mathrm{B}}, \tag{8.46}$$

$$\dot{x}_{\mathrm{p}} = \frac{Q_{\mathrm{B}}}{A_{\mathrm{B}}}. \tag{8.47}$$

Combining this with the orifice equation, piston velocity may be written as:

$$\dot{x}_{\mathrm{p}} = -\frac{C_{\mathrm{d}}\sqrt{\frac{2}{\rho}}}{A_{\mathrm{B}}} A_{\mathrm{d.B}}\sqrt{|p_{\mathrm{B}} - p_{\mathrm{S}}|} \tag{8.48}$$

if the pressure drop is negative ($(p_{\mathrm{S}} > p_{\mathrm{B}})$). The flow out of the piston on the A side is:

$$Q_{\mathrm{A}} = \dot{x}_{\mathrm{p}} A_{\mathrm{A}}. \tag{8.49}$$

With the adjustable throttle valve it is now possible to reset the flow area as the loading condition has changed. However, it is not feasible to adjust the throttle valve between every cycle if various loading conditions are experienced in each cycle. Note that the used valve setup requests the loading to be against the movement for the steady state "analysis" conducted to be true.

Using throttle valves for flow control, and indirectly for velocity control, may be simple and cheap. However, the fluid flow through the throttle valves depend on the pressure drop across the valves, i.e. if either supply or load pressure changes the flow also changes, leading to an error in the velocity control. To avoid the pressure dependency of the fluid flow, a pressure compensated flow control valve has been developed. In the following two sections we shall familiarise ourselves with two types of pressure compensated flow control valve.

8.4.3 *Pressure Compensated Flow Control Valve*

Examination of the orifice equation leads to the conclusion that a constant flow through a valve is secured by a constant pressure drop across a constant valve opening area. A constant flow is secured with a flow control valve consisting of a throttle valve combined with a pressure reduction valve. However, contrary to the pressure reduction valve in Fig. 8.26 the spring/pilot chamber is not connected to the tank but to the downstream pressure, p_{B} while the A port is connected to the upstream port of the throttle valve, see Fig. 8.33.

Note how the pressure reduction valve controls the pressure drop across the throttle valve to a preset value. The preset pressure difference is controlled by the spring force. If the pressure at A is high, compared to the pressure at B and the spring force, the pressure reduction valve (left valve) is closed a bit and the pressure at A will decrease. Conversely, if the pressure at B is high compared to pressure A, the pressure reduction valve (left valve) will open and the pressure at A will increase. The controlled pressure at A is limited by the supply pressure at P, i.e. if the load pressure B exceeds the pressure at P minus the preset pressure difference, a flow error will occur as the pressure drop is not as required.

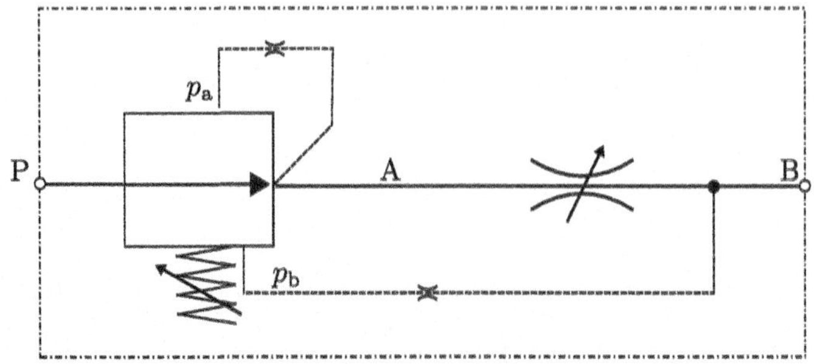

Fig. 8.33 Diagram of pressure compensated flow control valve

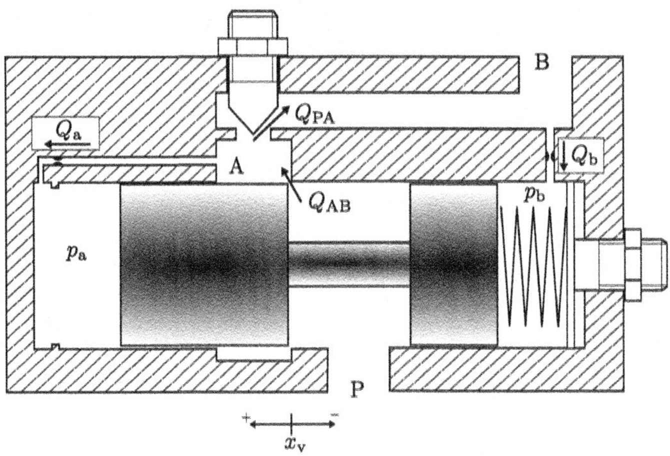

Fig. 8.34 Illustration of a pressure compensated flow control valve

An illustration of a pressure compensated flow control valve constructed in one valve block is seen in Fig. 8.34.

Note how the A chamber lives in between the throttle valve and pressure reduction valve orifices. Furthermore, it is worth noticing that small orifices are placed between pilot chambers and associated control volume (A→ a) and (B→ b).

Let's develop a model for the valve in Fig. 8.34 by setting up the describing equations. The notation follows those of the figure. The flows into the two pilot chambers are defined positive as is the flow from P to A and the flow from A to B. The four flows are:

$$Q_{\mathrm{PA}} = C_{\mathrm{d}} d_{\mathrm{v}} \pi x_{\mathrm{v}} \sqrt{\tfrac{2}{\rho} |p_{\mathrm{P}} - p_{\mathrm{A}}|} \operatorname{sgn}(p_{\mathrm{P}} - p_{\mathrm{A}}) (x_{\mathrm{v}} \geq 0), \tag{8.50}$$

$$Q_{\mathrm{AB}} = C_{\mathrm{d}} A_{\mathrm{d.AB}} \sqrt{\tfrac{2}{\rho} |p_{\mathrm{A}} - p_{\mathrm{B}}|} \operatorname{sgn}(p_{\mathrm{A}} - p_{\mathrm{B}}), \tag{8.51}$$

$$Q_{\mathrm{a}} = C_{\mathrm{d}} A_{\mathrm{d.a}} \sqrt{\tfrac{2}{\rho} |p_{\mathrm{A}} - p_{\mathrm{a}}|} \operatorname{sgn}(p_{\mathrm{A}} - p_{\mathrm{a}}), \tag{8.52}$$

$$Q_{\mathrm{b}} = C_{\mathrm{d}} A_{\mathrm{d.b}} \sqrt{\tfrac{2}{\rho} |p_{\mathrm{B}} - p_{\mathrm{b}}|} \operatorname{sgn}(p_{\mathrm{B}} - p_{\mathrm{b}}). \tag{8.53}$$

We see that only Q_{PA} is dependent on the spool position, x_{v}, of the pressure reduction valve. Three control volumes are defined for the valve while the P and B port pressures are seen as inputs to the valve model. The pressure gradient in the three control volumes are modelled as:

$$\dot{p}_{\mathrm{A}} = \tfrac{\beta}{V_{\mathrm{A}}} \left(Q_{\mathrm{PA}} - Q_{\mathrm{AB}} - Q_{\mathrm{a}} \right), \tag{8.54}$$

$$\dot{p}_{\mathrm{a}} = \tfrac{\beta}{V_{\mathrm{a0}} - x_{\mathrm{v}} A_{\mathrm{v}}} \left(Q_{\mathrm{a}} + \dot{x}_{\mathrm{v}} A_{\mathrm{v}} \right), \tag{8.55}$$

$$\dot{p}_{\mathrm{b}} = \tfrac{\beta}{V_{\mathrm{b0}} + x_{\mathrm{v}} A_{\mathrm{v}}} \left(Q_{\mathrm{b}} - \dot{x}_{\mathrm{v}} A_{\mathrm{v}} \right). \tag{8.56}$$

The movement of the spool in the pressure reduction part of the valve is modelled using Newton's Second Law:

$$\ddot{x}_{\mathrm{v}} = \frac{1}{m_{\mathrm{v}}} \left(A_{\mathrm{v}} (p_{\mathrm{b}} - p_{\mathrm{a}}) + k_{\mathrm{s}} (x_{\mathrm{v0}} - x_{\mathrm{v}}) - F_{\mathrm{fr}}(\dot{x}_{\mathrm{v}}) + F_{\mathrm{fl}} \right), \tag{8.57}$$

where m_{v} is the spool mass, k_{s} is the spring constant, F_{fl} is a function for the flow force and $F_{\mathrm{fr}}(\dot{x}_{\mathrm{v}})$ is a friction model dependent on the spool velocity $\dot{x}_{\mathrm{v}}$.

The model developed is rather simple as e.g. no leakage flow is included.

Steady State Model

Often the eigenfrequency of the pressure compensator valve is so high that the steady state equations are feasible in context of system modelling. However, we shall investigate the influence of various components on system models later.

$$Q_{\mathrm{PA}} = C_{\mathrm{d}} d_{\mathrm{v}} \pi x_{\mathrm{v}} \sqrt{\tfrac{2}{\rho} |p_{\mathrm{P}} - p_{\mathrm{A}}|} \operatorname{sgn}(p_{\mathrm{P}} - p_{\mathrm{A}}) (x_{\mathrm{v}} \geq 0), \tag{8.58}$$

$$Q_{\mathrm{AB}} = C_{\mathrm{d}} A_{\mathrm{d.AB}} \sqrt{\tfrac{2}{\rho} |p_{\mathrm{A}} - p_{\mathrm{B}}|} \operatorname{sgn}(p_{\mathrm{A}} - p_{\mathrm{B}}), \tag{8.59}$$

$$Q_{\mathrm{PA}} = Q_{\mathrm{AB}}, \tag{8.60}$$

$$0 = A_{\mathrm{v}} (p_{\mathrm{B}} - p_{\mathrm{A}}) + k_{\mathrm{s}} (x_{\mathrm{v0}} - x_{\mathrm{v}}) + F_{\mathrm{fl}}, \tag{8.61}$$

Note that the pressure compensator of the current valve controls the internal pressure at point A relative to the pressure at port B.

8.4.4 Pressure Compensated Flow Control Valve—Bypass

The pressure compensated flow control valve with bypass controls the pressure drop across the fixed orifice to be constant, by passing excess flow directly to the tank rather than just blocking it out. As we shall discuss soon the two pressure compensated flow control valves' topologies affect the inlet pressure rather differently. Figure 8.35 shows the diagram for a bypass pressure compensated flow control valve. The pressure compensator valve is merely a pressure relief valve with pilot chamber b connected to the back connected to port B.

Figure 8.36 illustrates how a pressure compensated flow control valve with bypass may be constructed in a valve block. Note that small orifices are fitted in the pilot lines. It can be seen how both the pressure preset and the throttle valve opening area may be adjusted with a screw.

Let's set up the describing equations for the pressure compensated flow control valve with bypass. We start with four orifice equations for the flows illustrated in the figure,

$$Q_{\mathrm{AT}} = C_{\mathrm{d}} d_{\mathrm{v}} \pi x_{\mathrm{v}} \sqrt{\frac{2}{\rho} |p_{\mathrm{A}} - p_{\mathrm{T}}|} \operatorname{sgn}(p_{\mathrm{A}} - p_{\mathrm{T}}) (x_{\mathrm{v}} \geq 0), \tag{8.62}$$

$$Q_{\mathrm{AB}} = C_{\mathrm{d}} A_{\mathrm{d.AB}} \sqrt{\frac{2}{\rho} |p_{\mathrm{A}} - p_{\mathrm{B}}|} \operatorname{sgn}(p_{\mathrm{A}} - p_{\mathrm{B}}), \tag{8.63}$$

$$Q_{\mathrm{a}} = C_{\mathrm{d}} A_{\mathrm{d.a}} \sqrt{\frac{2}{\rho} |p_{\mathrm{A}} - p_{\mathrm{a}}|} \operatorname{sgn}(p_{\mathrm{A}} - p_{\mathrm{a}}), \tag{8.64}$$

$$Q_{\mathrm{b}} = C_{\mathrm{d}} A_{\mathrm{d.b}} \sqrt{\frac{2}{\rho} |p_{\mathrm{B}} - p_{\mathrm{b}}|} \operatorname{sgn}(p_{\mathrm{B}} - p_{\mathrm{b}}), \tag{8.65}$$

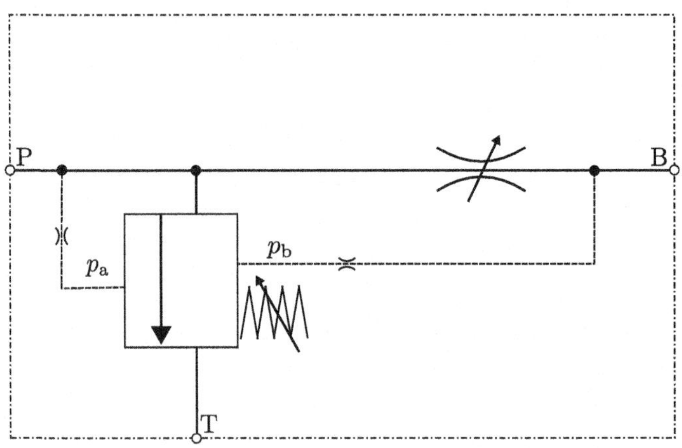

Fig. 8.35 Fluid power diagram of a pressure compensated flow valve with flow bypass

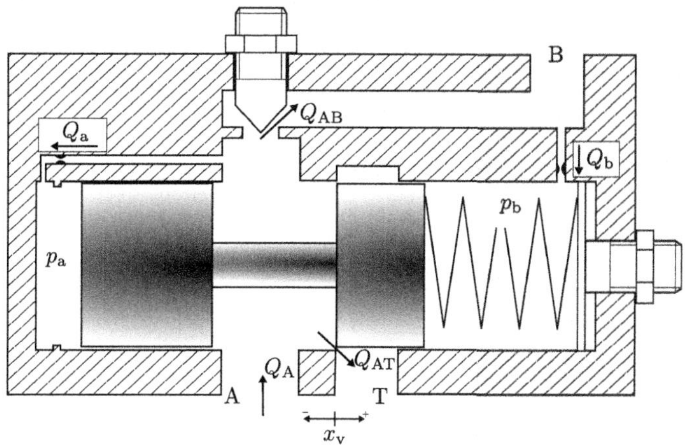

Fig. 8.36 Illustration of a pressure compensated flow valve with flow bypass

and we note that the flow, Q_A, into the valve is:

$$Q_A = Q_{AB} + Q_{AT} + Q_a. \tag{8.66}$$

Three control volumes and flow continuity equations of interest are defined for the valve block as:

$$\dot{p}_A = \frac{\beta}{V_A}\left(Q_A - Q_{AB} - Q_{AT} - Q_a\right), \tag{8.67}$$

$$\dot{p}_a = \frac{\beta}{V_{a0} + x_v A_v}\left(Q_a - \dot{x}_v A_v\right), \tag{8.68}$$

$$\dot{p}_b = \frac{\beta}{V_{b0} - x_v A_v}\left(Q_b + \dot{x}_v A_v\right). \tag{8.69}$$

The spool movement does not influence the volume of chamber A as the left and right wall of the control volume moves synchronously. Lastly we setup Newton's Second Law for the spool:

$$\ddot{x}_v = \frac{1}{m_v}\left(A_v(p_a - p_b) - k_s(x_{v0} + x_v) - F_{fr}(\dot{x}_v) + F_{fl}\right) \tag{8.70}$$

Steady State Model

Once again let's construct a steady state model:

$$Q_{AT} = C_d d_v \pi x_v \sqrt{\frac{2}{\rho}|p_A - p_T|}\operatorname{sgn}(p_A - p_T)(x_v \geq 0), \tag{8.71}$$

$$Q_{AB} = C_d A_{d.AB} \sqrt{\frac{2}{\rho}|p_A - p_B|}\operatorname{sgn}(p_A - p_B), \tag{8.72}$$

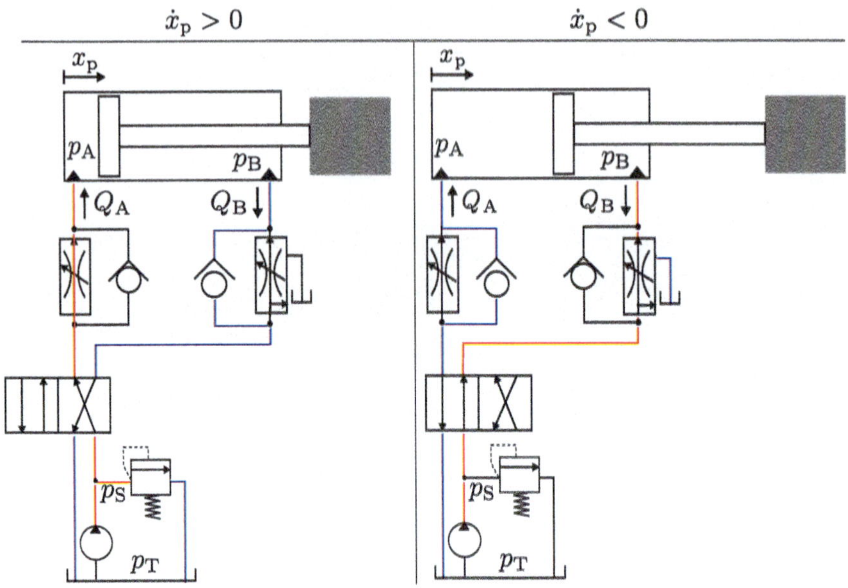

Fig. 8.37 Diagram of differential cylinder velocity control with pressure compensated flow control valves

$$Q_{AB} = (Q_A - Q_{AT})\,. \tag{8.73}$$

$$0 = A_v(p_A - p_B) - k_s(x_{v0} + x_v) + F_{fl} \tag{8.74}$$

Note that the pressure compensator for the current valve controls the pressure at port A, hence the inlet port pressure is controlled. As inlet to the valve, port A is likely connected to other components whose inlet pressure is now controlled by the flow control valve.

Case Illustration—Pressure Compensated Flow Control Valve

Let's examine a case where pressure compensated flow control valves are used for velocity control of a differential cylinder. The case setup is designed as metering in with pressure compensated flow control valves and return check valves as seen in Fig. 8.37. The coloured lines indicate pressure in the two situation, extending and retracting the piston (left and right).

Note that the flow rate is controlled in the inlet path, hence metering in is applied in both directions. The A side chamber utilises a standard pressure compensated flow control valve, whereas the B side chamber uses a pressure compensated flow control valve with bypass. The supply pressure is as such controlled by the pressure relief valve when extending the piston and the bypass flow control valve when retracting the piston.

We shall accept that the metering in control is only feasible when the load force is opposing movement in both directions, hence no overrunning load may be controlled with this setup. Consider what will happen if the pump system present in Fig. 8.37 is to supply fluid for multiple numbers of flow control valves—which valve type would you choose?

As the bypass valve type affects the supply pressure, one must be careful to utilise this type in parallel with other consumers.

Extending Piston

The flow through the flow control valve is set with the opening area $A_{0.A}$ and is modelled as:

$$Q_A = C_d A_{0.A} \sqrt{\frac{2}{\rho} \Delta p_{set}}. \tag{8.75}$$

Assuming steady state the piston velocity $\dot{x}_p$ and out flow Q_B are given as:

$$Q_A = \dot{x}_p A_A \quad \Rightarrow \quad \dot{x}_p = \frac{Q_A}{A_A}, \tag{8.76}$$

$$Q_B = \dot{x}_p A_B \quad \Rightarrow \quad \dot{x}_p = \frac{Q_B}{A_B}. \tag{8.77}$$

In conclusion the velocity is;

$$\dot{x}_p = \frac{C_d \sqrt{\frac{2}{\rho}}}{A_A} A_{0.A} \sqrt{\Delta p_{set}}, \tag{8.78}$$

and the pump pressure will be controlled by the pressure relief valve to, $p_S = p_{cr}$ if $Q_P > Q_A$.

Retracting Piston

The flow through the flow control valve is set with the opening area $A_{0.B}$ and is modelled as:

$$Q_B = C_d A_{0.B} \sqrt{\frac{2}{\rho} |\Delta p_{set}|}. \tag{8.79}$$

Assuming steady state the piston velocity $\dot{x}_p$ and out flow Q_A are given as:

$$-Q_B = -\dot{x}_p A_B \quad \Rightarrow \quad \dot{x}_p = \frac{Q_B}{A_B}, \tag{8.80}$$

$$Q_A = \dot{x}_p A_A \quad \Rightarrow \quad \dot{x}_p = \frac{Q_A}{A_A}. \tag{8.81}$$

In conclusion the velocity is;

$$\dot{x}_\mathrm{p} = -\frac{C_\mathrm{d}\sqrt{\frac{2}{\rho}}}{A_\mathrm{B}} A_{0.\mathrm{B}}\sqrt{\Delta p_\mathrm{set}}, \tag{8.82}$$

and the pump pressure will be given as:

$$p_\mathrm{S} = p_\mathrm{B} + \Delta p_\mathrm{set} + \Delta p_\mathrm{d}, \tag{8.83}$$

where Δp_d and Δp_set is the pressure drop across the directional valve and the pressure drop in the flow control valve respectively. Furthermore, the supply pressure is requested to be lower than the cracking pressure of the relief valve $p_\mathrm{S} < p_\mathrm{cr}$.

8.5 Pressure Compensated Proportional Valves

Pressure compensated proportional valves are proportional valves which are pressure compensated such that the pressure drop across the main spool in the flow path from supply pressure is held constant. In other words, the pressure before the main spool is controlled with a pressure reduction valve, controlled with pilot pressure from either main spool pressure in- or outlet port depending on the valve state, see Fig. 8.38.

Note that one pilot pressure is always the main spool P connection and the other pilot pressure is either port A or B depending on valve state a or c, respectively.

With the spring the pressure compensator valve is preset to keep a constant pressure drop, such that the flow through the pressure compensated proportional valve may be modelled (when the proportional valve is in state a) as:

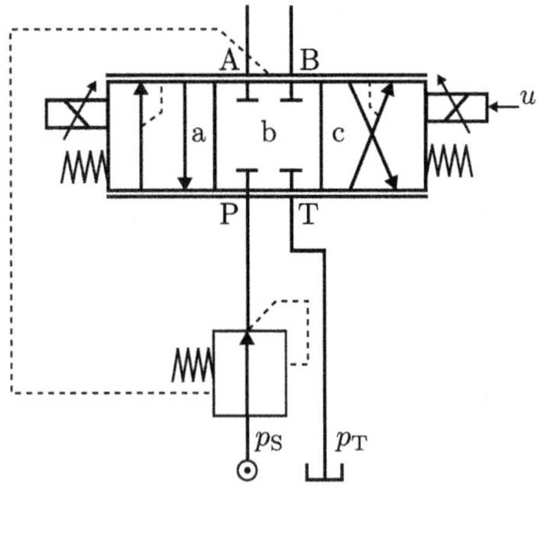

Fig. 8.38 Diagram of differential cylinder velocity control with pressure compensated flow control valves

$$Q(x_v, p_P, p_A) = C_d A_d(x_v)\sqrt{\frac{2}{\rho}|p_P - p_A|}\text{sgn}(p_P - p_A). \tag{8.84}$$

Note that the pressure drop $p_P - p_A$ is constant Δp due to the pressure reduction valve. This enforces that flow through the valve is only a function of the opening determined by the valve spool position:

$$Q(x_v) = C_d A_d(x_v)\sqrt{\frac{2}{\rho}\Delta p} = k_v A_d(x_v). \tag{8.85}$$

We have seen that the flow is dependent on valve coefficients and the spool position x_v. When used for velocity control of a fluid power actuator the steady state velocity becomes proportional to the valve opening (spool position for a linear opening area function).

References

1. Check valve, size 04 ... 16, RKVE series, Reference: 170-P-051300-EN-04, issue: 08.2018. Bucher Hydraulics Dachau GmbH
2. 2/2 Cartridge seat valve, WS22DG series, Reference: 400-P-121110-EN-00, issue: 09.2015. Bucher Hydraulics AG Frütigen
3. Servovalves direct drive servovalves D633/634 series, Rev 2. 04/2009. Moog Inc
4. Servo valves pilot operated - flow control valve with analog interface, G761/-761 series, Rev L (2021)
5. Catalog HY15-3502/US, Direct acting relief valve, series A04B2. Parker Hannifin Corporation
6. Throttle and shut-off cartridges, size 5. Reference: 400-P-405071-EN-01, issue: 05.2020, Bucher Hydraulics AG Frütigen

Chapter 9
Accumulators

Abstract This chamber introduces very basic accumulator models for a mass loaded, a spring-loaded and a gas-loaded accumulator.

Accumulators have two major functions in fluid power systems: firstly, accumulators are used to stabilise pressure; secondly, accumulators are used as energy storage. So accumulators are for fluid power systems what capacitors are for electrical systems.

Accumulators are constructed in various ways and with different means of energy accumulation. In Fig. 9.1 five accumulator types are illustrated. One can see three types of energy accumulation: mass, mechanical spring and compressed gas. Three types of gas type accumulators are also seen.

In modern fluid power systems gas accumulators are the most commonly used. Each of the three gas type accumulators are used as each construction has pros and cons for different systems. In the next sections, various accumulator models are presented to familiarise the reader with accumulators. This note, however, will not go deeply into accumulator models. Interested readers are asked to refer to more dedicated material on accumulators.

9.1 Piston Accumulator

The first three accumulators depicted in Fig. 9.1 from left to right are piston type accumulators, however, with different means of energy accumulation. The various topologies yield different pressure behaviours for the fluid in the accumulators.

9.1.1 Mass Loaded Piston Accumulators

Mass load accumulators are seldom used as the size of the mass required is often large compared to the system under consideration. A model for the mass accumulator is simply like a one chamber cylinder carrying a gravitational load M_L:

A. H. Hansen, *Fluid Power Systems*, Fluid Mechanics and Its Applications 129,
https://doi.org/10.1007/978-3-031-15089-0_9

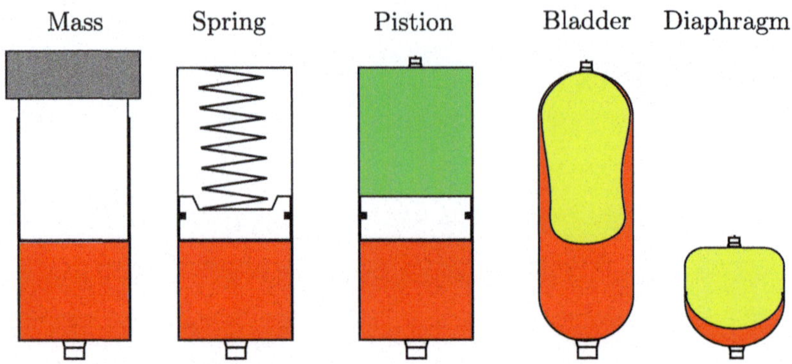

Fig. 9.1 Illustration of accumulator types

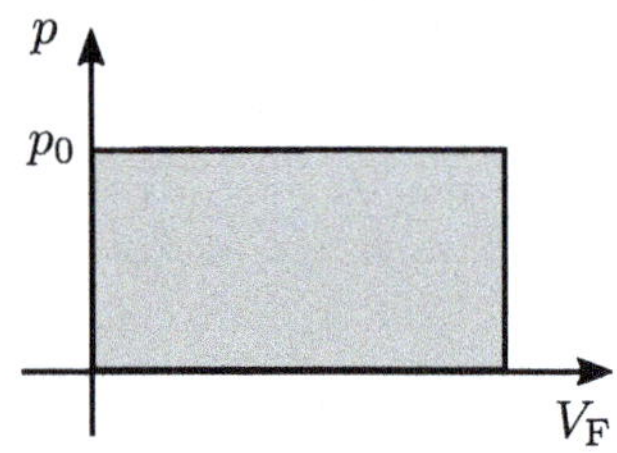

Fig. 9.2 Illustration of pressure diagram for mass loaded accumulator

$$\ddot{x}_p M_L = A p_f - F_{fr}(\dot{x}_p) - M_L g, \tag{9.1}$$

$$\dot{p}_f = \frac{\beta(p_f)}{V_f(x_p)} \left(Q_{acc} - \dot{x}_p A \right), \tag{9.2}$$

$$V_f(x_p) = V_{f0} + A x_p, \tag{9.3}$$

where x_p and A are the piston position and area respectively, $F_{fr}(\dot{x}_p)$ is the friction model and $\frac{\beta(p_f)}{V_f(x_p)}$ is the ratio between bulk modulus of the fluid and fluid volume.

As the mass to be moved is large the dynamic behaviour of the accumulator may be slow. However, if assumed negligible the steady state model entails that the fluid pressure is constant if the accumulator is not in either of the end position. The steady state model is such given as:

$$p_f = \frac{M_L g}{A}, \tag{9.4}$$

$$V_f = V_{f0} + A x_p. \tag{9.5}$$

An illustration of the volume, pressure diagram is seen in Fig. 9.2.

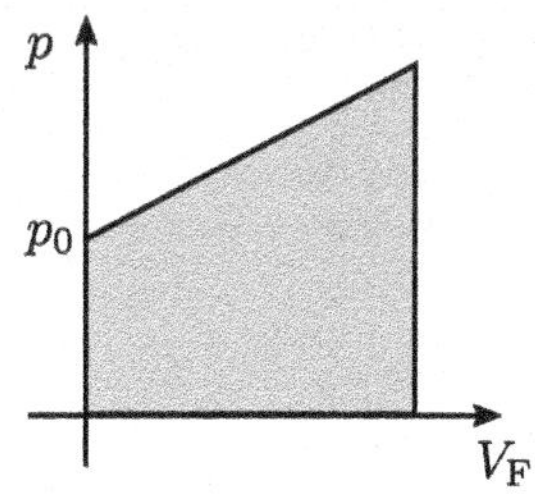

Fig. 9.3 Pressure illustration for spring loaded piston accumulator

9.1.2 *Spring Loaded Piston Accumulators*

A simple model excluding thermal effects and assuming the spring chamber to be vented to its surrounding is given as:

$$\ddot{x}_p M_p = A p_f - F_{fr}(\dot{x}_p) - k_s(x_p + x_{p0}), \tag{9.6}$$

$$\dot{p}_f = \frac{\beta(p_f)}{V_f(x_p)} \left(Q_{acc} - \dot{x}_p A\right), \tag{9.7}$$

$$V_f(x_p) = V_{f0} + A x_p, \tag{9.8}$$

where the notation follows the mass loaded accumulator with the addition of the spring stiffness, k_s.

If one may assume the piston movement to be significantly faster than the pulsations in the accumulator flow, one may use a steady state model which reads:

$$0 = A p_f - k_s(x_p + x_{p0}), \tag{9.9}$$

$$V_f = V_{f0} + A x_p. \tag{9.10}$$

When using the steady state model the pressure in the fluid in the accumulator may be depicted as in Fig. 9.3.

Note that the pressure is $p_0 - \frac{k_s x_{p0}}{A}$ as the first fluid enters and how the pressure increases linearly with increasing fluid volume. This is true when assuming that the spring is only operating in the linear part and a spring pre-compression of x_{p0} is employed.

9.1.3 *Gas Loaded Piston Accumulators*

Modeling a gas loaded piston accumulator is like modeling a cylinder with no external load, though differing by the pressure model for the gas chamber pressure. The describing equations for the piston accumulator are given as:

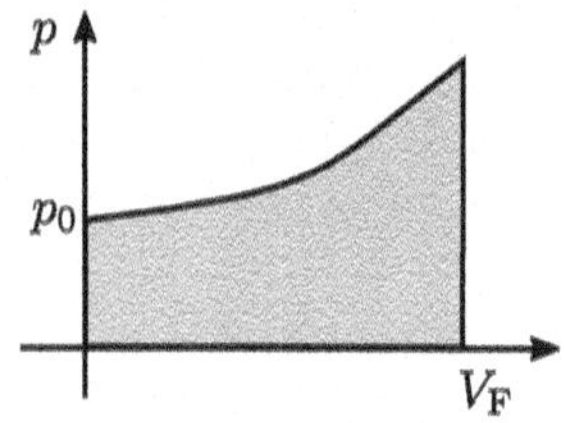

Fig. 9.4 Illustration of a static pressure curve for a gas loaded piston accumulator

$$\ddot{x}_p M_p = A p_f - A p_g - F_{fr}(\dot{x}_p), \tag{9.11}$$
$$\dot{p}_f = \frac{\beta(p_f)}{V_f(x_p)} \left(Q_{acc} - \dot{x}_p A\right), \tag{9.12}$$
$$V_f = V_{f0} + A x_p. \tag{9.13}$$

The gas pressure model is dependent on the operation cycles of the accumulator which entails the value of the poly-tropic index n ranging from 1 for an isothermal process to 1.4 for an adiabatic process. We shall use the following gas law:

$$p_g V_g^n = \text{constant}, \tag{9.14}$$
$$V_g = V_{g0} - A x_p, \tag{9.15}$$
$$p_g = \frac{p_{g0} V_{g0}^n}{V_g^n}. \tag{9.16}$$

Gas in the upper chamber instead of a spring leads to a non-linear "spring" force as the stiffness of a gas is increased as pressure increases. The relation between fluid in the accumulator and pressure in said fluid is given in Fig. 9.4. Compressing the model to two differential equations with the accumulator in flow as input yields:

$$\ddot{x}_p M_p = A p_f - A \frac{p_{g0} V_{g0}^n}{(V_{g0} - A x_p)^n} - F_{fr}(\dot{x}_p), \tag{9.17}$$
$$\dot{p}_f = \frac{\beta(p_f)}{V_{f0} + A x_p} \left(Q_{acc} - \dot{x}_p A\right). \tag{9.18}$$

9.2 Bladder Accumulator

In a bladder accumulator the fluid and gas are separated with a rubber bladder. The fluid is in a steel container and the gas is in a bladder inside the steel container (the container may be of other materials, e.g. carbon fibre). With the very low mass of the bladder, pressure in the gas and fluid is practically equal $p_f = p_g$. One may use a flow continuity equation to formulate the pressure gradient in the fluid and a gas law to calculate the fluid volume under a given pressure situation:

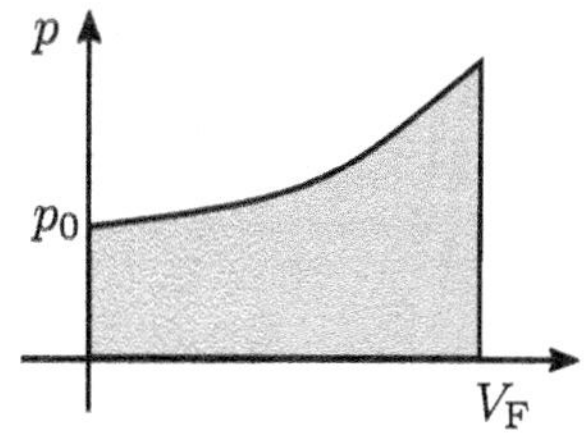

Fig. 9.5 Illustration of a static pressure curve for a gas loaded bladder accumulator

$$\dot{p}_\text{f} = \frac{\beta(p_\text{f})}{V_\text{f}} \left(Q_\text{acc} - \dot{V}_\text{f} \right), \tag{9.19}$$

$$p_\text{g} = p_\text{f}(p_\text{f} \geq p_\text{g0}) + p_\text{g0}(p_\text{f} < p_\text{g0}) \tag{9.20}$$

Note that the gas pressure is equal to the fluid pressure when the fluid pressure is higher than the pre-change pressure. The fluid volume is given as the difference in total accumulator volume and gas volume:

$$V_\text{f} = V_\text{g0} - V_\text{g}, \tag{9.21}$$

$$V_\text{g} = \left(\frac{p_\text{g0} V_\text{g0}^n}{p_\text{g}} \right)^{\frac{1}{n}}. \tag{9.22}$$

The change from a piston to a bladder gas accumulator is primarily seen in the dynamic behaviour and the structural differences. Therefore static pressure curves for the piston and bladder type gas accumulators are similar (Fig. 9.5).

9.3 Diaphragm Accumulator

The diaphragm accumulator is modeled as the bladder accumulator. The two accumulators differ in the way the gas and fluid chambers are separated. The bladder accumulator leads to a larger workable difference volume.

Chapter 10
Fluid Power Transmission Lines

Abstract This chapter introduces a lumped parameter time-domain model for fluid power transmission lines. Before the modelling part, the pressure wave speed in the working fluid relative to the line length is discussed to evaluate if a dynamic or steady model is most reasonable.

Pipes and hoses are used to transport pressurised fluid in fluid power transmission systems. Often pumps and actuators are located some distance from each another, requiring pipes or hoses to transport the fluid from the pumps to the actuators, possibly with the valves inbetween controlling flow direction, amount and pressure. With pipes and hoses, the transmission of energy from one point to another may be done with a huge spacial freedom compared to a purely mechanical transmission.

In this chapter we shall briefly present the physical pipe and hose construction and familiarise ourselves with how the pipe and hose dimension may be chosen. However, as this lecture note focuses on system modelling and analysis, more dedicated material may be consulted if hose and pipe types, fittings and the construction of these are of interest.

Modelling of fluid transmission lines (pipe/hoses) will be discussed both for dynamic models and steady state models. The choice of transmission line model is highly dependent on the purpose of the modelled and the operating condition of the system. With regard to the latter, one may illustrate this by the simple system seen in Fig. 10.1, where a fluid is flowing from a pressure source through a fixed orifice, a pipe and an on/off valve. When the valve is suddenly closed the flow is blocked and a pressure increase will happen in front of the valve.

As the flow passage to the tank is blocked, flow in the hose must be stopped. However, this does not happen instantaneously along the length of the hose, but from the closed valve towards the inflow. The pressure increase at the valve will induce a force opposing the flow direction and thus decelerate the fluid flow. The travel time of this pressure wave is finite as the pressure wave velocity is given from the fluid compressibility as the speed of sound, c, in the fluid,

$$c = \sqrt{\frac{\beta}{\rho}}, \tag{10.1}$$

A. H. Hansen, *Fluid Power Systems*, Fluid Mechanics and Its Applications 129,
https://doi.org/10.1007/978-3-031-15089-0_10

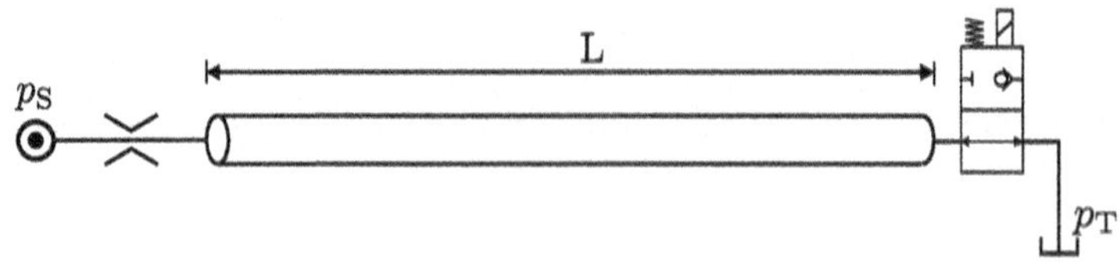

Fig. 10.1 Illustration of a transmission line and valve connecting supply and tank pressure

where β and ρ are bulk modulus and density of the oil respectively. The round trip time, T, for a pressure wave travelling from one end of a hose with length, L, to the other and back is given as,

$$T = \frac{2L}{c}. \tag{10.2}$$

With a fluid bulk modulus of 16000 bar, a density of 860 $\frac{kg}{m^3}$ and a hose length of 1m the speed of the pressure wave and round trip time are:

$$c = \sqrt{\frac{\beta}{\rho}} = \sqrt{\frac{16000 \cdot 10^5 \text{Pa}}{860 \frac{\text{kg}}{\text{m}^3}}} = 1364 \frac{\text{m}}{\text{s}}, \tag{10.3}$$

$$T = \frac{2 \cdot 1 \text{ m}}{1364 \frac{\text{m}}{\text{s}}} = 0.00147\text{s}. \tag{10.4}$$

If the valve is slowly closed decreasing the flow slowly to zero, the flow will continuously be decelerated avoiding the high pressure peak at the valve leading to the pressure wave. But what is slow? In the current context the valve closing time T_c (how much time is used for the flow change at the valve) should be compared to the pressure wave round trip time. If the flow changes slowly ($T_c << T$) a steady state transmission line model is adequate.

However the system excitation T_c may be fast compared to the hose dynamics, requesting a transmission line model including pressure dynamics in the fluid lines. Once again you as an engineer have to choose the appropriate level of model complexity

10.1 Steady State Transmission Line Model

In steady state models, fluid transmission lines are modelled using the flow/pressure relations described in Sect. 4.2, "Flow in a Pipe", where a given flow induces a corresponding pressure loss. This model assumes the pressure development in the hose to be infinitely fast, which is why this model requests the hose inflow to vary slowly compared to the pressure wave round trip time.

10.2 Dynamic Transmission Line Model

When the excitation at one end of the transmission line is fast the request for a dynamic transmission line model may arise. Various model approaches may be used in modelling the pressure and flow dynamics in fluid transmission lines.

Frequency based models in the Laplace domain and lumped parameter models in the time domain are two well known approaches. In the next section we shall introduce a simple lumped parameter model.

10.2.1 Lumped Parameter Model

In lumped parameter modelling of fluid transmission lines the line is separated in n number of hose elements as illustrated in Fig. 10.2. In each element fluid parameters are lumped to one value, such that pressure, density, bulk modulus, viscosity and temperature are constant in spacial coordinates in that element.

The lumped parameter transmission line model is built from Newton's Second Law and the flow continuity equation. Following notation from Fig. 10.2 Q_i is the flow into element i in which the pressure is p_i. The acceleration of the mass, m_i of element i is derived as:

$$\ddot{x}_{im_i} = (p_{i-1} - p_i - \Delta p_{\mathrm{fr},i})A_{\mathrm{n}}, \tag{10.5}$$

$$\ddot{x}_i = \frac{(p_{i-1} - p_i - \Delta p_{\mathrm{fr},i}}{m_i})A_{\mathrm{n}}, \tag{10.6}$$

$$\ddot{x}_i = \frac{(p_{i-1} - p_i) - \Delta p_{\mathrm{fr},i})A_{\mathrm{n}}}{L_{\mathrm{n}} A_{\mathrm{n}} \rho_i}, \tag{10.7}$$

where L_{n} and A_{n} are hose element length and cross section area respectively, ρ_i is oil density in the i'th element while $\Delta p_{\mathrm{fr},i}$ is the flow induced pressure loss in the element. In fluid power systems we often use volume flow rather than mass flow such that the change in volume flow is given as:

$$\ddot{x}_i A_{\mathrm{n}} = \frac{(p_{i-1} - p_i - \Delta, p_{\mathrm{fr},i})A_{\mathrm{n}}}{L_{\mathrm{n}} A_{\mathrm{n}} \rho} A_{\mathrm{n}}, \tag{10.8}$$

$$\dot{Q}_i = \frac{A_{\mathrm{n}}(p_{i-1} - p_i - \Delta p_{\mathrm{fr},i})}{\rho_i L_{\mathrm{n}}}. \tag{10.9}$$

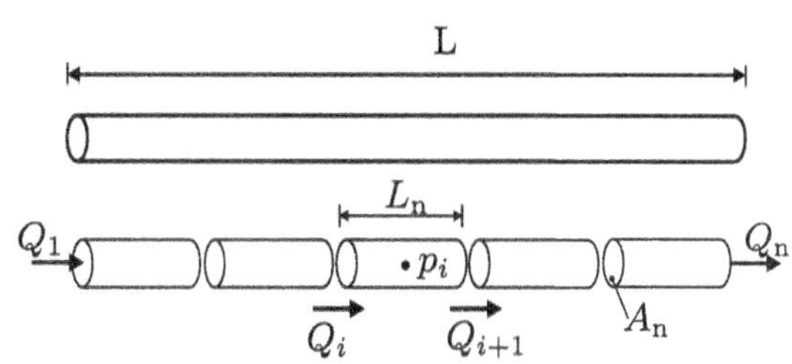

Fig. 10.2 Illustration of a transmission line and the modelling notation

The pressure gradient in each element is modelled with the flow continuity equation:

$$Q_i - Q_{i+1} = \dot{V}_n + \frac{V_n}{\beta_i}\dot{p}_i, \tag{10.10}$$

$$\dot{p}_i = \frac{\beta_i}{V_n}\left(Q_i - Q_{i+1} - \dot{V}_n\right), \tag{10.11}$$

where β_i is the bulk modulus of the oil element i and V_n is the volume of each hose element. Note that we use A_n and V_n as we assume the cross sectional flow area and the volume of each element to be equal. As such we divide the hose or pipe in n equal elements.

In this model we shall use a rather simple model for pressure drop due to fluid friction $\Delta p_{fr,i}$, which builds on the model developed in Sect. 4.2 for pressure loss in pipes due to fully laminar or fully turbulent flow. The pressure loss through one element is given as:

$$\Delta p_{lam}, i = \frac{128\mu L_n}{\pi d^4} Q_i, \tag{10.12}$$

$$\Delta p_{tur}, i = 0.242\frac{\mu^{0.25}\rho_i^{0.75} L_n}{d^{4.75}} Q_i^{1.75}, \tag{10.13}$$

where μ is the dynamic oil viscosity. In our time domain model we like a smooth transition between the laminar and turbulent friction models. Therefore, a hyperbolic tangent function is used with Reynolds number such that the combined pressure loss model is given as:

$$\Delta p_{fr,i} = \Delta p_{lam,i}\frac{1 - \tanh\left(\frac{Re-2150}{50}\right)}{2} + \Delta p_{tur,i}\frac{\tanh\left(\frac{Re-2150}{50}\right) + 1}{2} \tag{10.14}$$

The friction model is illustrated in Fig. 10.3 with plots for the combined pressure drop, the individual turbulent and laminar pressure drop and Reynolds number.

The lumped parameter model can be seen in the box below. Note that the hose inflow (Q_1) and the hose outflow (Q_n) are inputs to the model.

$$\dot{p}_i = \frac{\beta_i}{V_n}\left(Q_i - Q_{i+1} - \dot{V}_n\right) \tag{10.15}$$

$$\dot{Q}_i = \frac{A_n(p_{i-1} - p_i - \Delta p_{fr,i})}{\rho_i L_n} \tag{10.16}$$

$$\Delta p_{lam,i} = \frac{128\mu L_n}{\pi d^4} Q_i \tag{10.17}$$

$$\Delta p_{tur,i} = 0.242\frac{\mu^{0.25}\rho_i^{0.75} L_n}{d^{4.75}} Q_i^{1.75} \tag{10.18}$$

$$\Delta p_{fr,i} = \Delta p_{lam,i}\frac{1-\tanh\left(\frac{Re-2150}{50}\right)}{2} + \Delta p_{tur,i}\frac{\tanh\left(\frac{Re-2150}{50}\right)+1}{2} \tag{10.19}$$

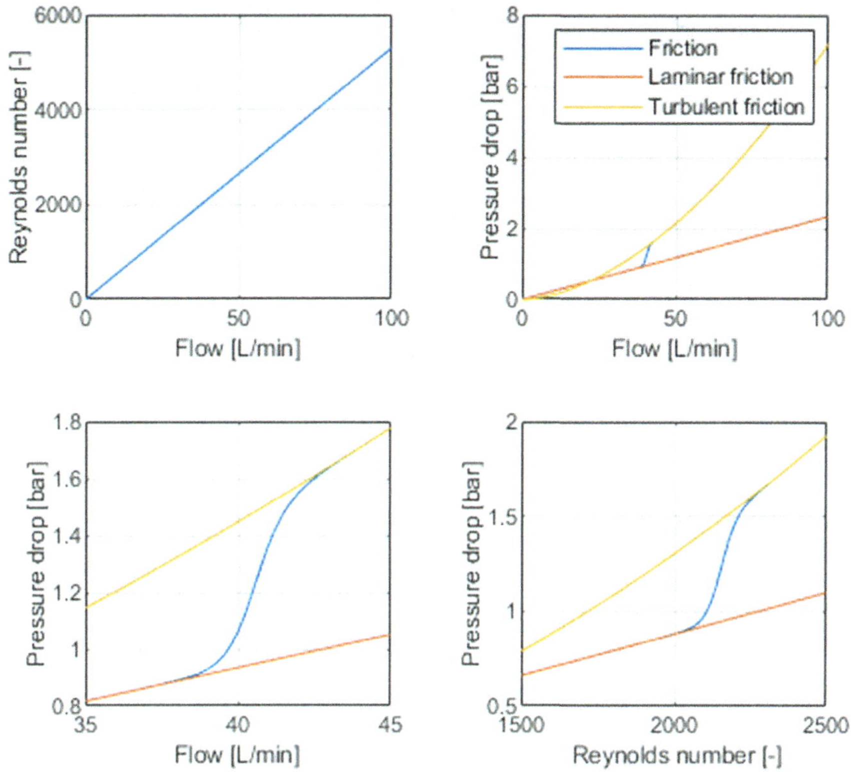

Fig. 10.3 Illustration of the friction model used in the lumped parameter transmission line model, $d = 10$ mm, $L_n = 1$ m, $\rho = 900$ kg/m^3, $\mu = 40 \cdot 10^{-6} \cdot 860 \frac{\mathrm{Ns}}{\mathrm{m}^2}$

10.3 Fluid Power Pipes and Hoses

Steel pipes and rubber-steel hoses are used as fluid transmission lines in fluid power systems. Due to safety and wear it is preferred, where possible, to use steel piping. Machines featuring moving parts may call for flexible hose connections as e.g. on an excavator's arm where the actuator and the HPU are moving relative to each other. For both pipes and hoses some restriction of bending radius exists and must be considered in the physical design. Readers should refer to hose and pipe suppliers to get information on diameter, allowable pressure level, bending radius etc.

10.3.1 Construction of Hoses

Fluid power hoses are made of layers of various materials. An inner layer is designed to be resistant to various oil types and additives, such that the oil does not degrade the hose. Then there is a middle section consisting of a number of steel layers separated by rubber layers to make the hose flexible. A larger diameter calls for a stronger hose requiring more layers of steel when the allowable operation pressure is held at e.g. 420 bar.

Part III
Fluid Power Systems

Chapter 11
Modelling Fluid Power Systems

Abstract This chapter introduces modeling of fluid power systems. Two "case" systems, in the form of a motor-valve drive and a cylinder-valve drive, are used to exemplify, firstly, how a lumped parameter time domain model is developed, and secondly, how a linear time domain model is derived from the former model. The chapter includes an illustration of how to solve the numerical time domain model by simulation. Furthermore, the chapter includes a linear recap time-invariant system, change variables and choice of linearization point.

In the current chapter we shall study how to model fluid power systems. Firstly, we discuss which type of models are required for various problems. Secondly, models are developed for some typical example systems, e.g. a simple cylinder-valve drive and a valve controlled motor. Lastly some comments and tips for setting up and debugging your numerical simulation tools round off the chapter.

11.1 Models

As evident in the fluid mechanic part, multiple methods exist when describing the physics of fluid, e.g. differential approach describing fluid parameters in a very small volume of fluid; ($dV = dxdydz$) and the control volume approach where the fluid parameters are said to be equal in a larger control volume (lumped parameters). When choosing which model type to adopt the engineer needs to consider the purpose of modelling a given system. Depending on the question being asked and the problem to be solved a model type is chosen accordingly. When a position controller is to be designed for the cylinder piston in the system in Fig. 11.1, lumped parameter modelling is chosen. However, if the question is how the power loss across the 4/3 way valve may be decreased by redesigning flow paths in the valve, a finite volume model may be required to analyse the flow pattern and the pressure losses. However, for such an analysis it is infeasible to construct a CFD model of the whole system; instead a CFD model for the valve itself may be derived where the flow or pressure boundaries may be taken from the lumped parameter system model.

A. H. Hansen, *Fluid Power Systems*, Fluid Mechanics and Its Applications 129,
https://doi.org/10.1007/978-3-031-15089-0_11

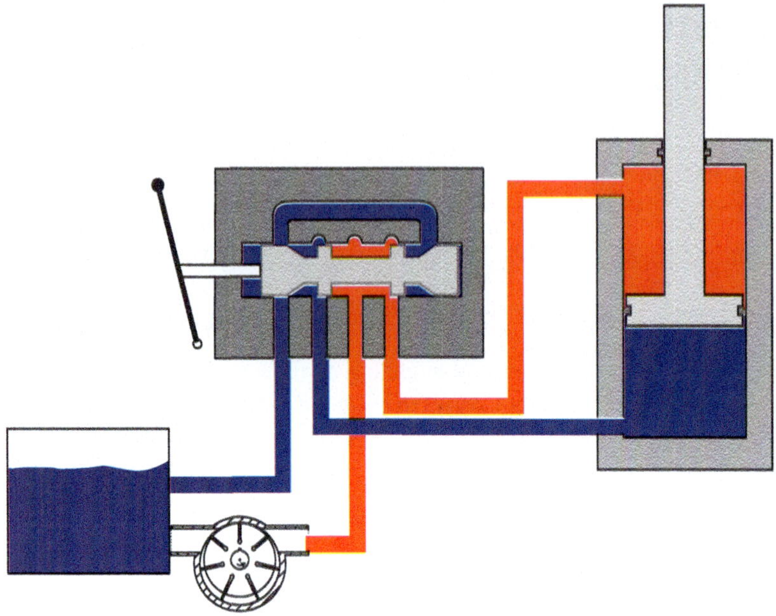

Fig. 11.1 System illustration

The lecture note at hand focuses on lumped parameter modelling of components and systems.

11.1.1 Time Domain Model—Non-linear

In time domain models the focus is on modelling how system "states" develop as time passes. The states of a system are the time varying variables included in the model, e.g. position, velocity, acceleration may be states of a system that change as time passes, especially if an external force is applied. In fluid power systems, pressure, flow and temperature may be states that change with time. As modelled constructor, however, you decide what the states are in your model. Even though a system's "state" is most certainly changing with time one may chose to excluded it from the model. In later case studies we will mostly exclude temperature and even flow changes. One may refer back to hose modelling: when adopting a steady model the flow change is unmodelled, while it is included when using the lumped parameter transmission model.

When developing lumped parameter models for systems and components we mostly adopt position and velocity of moving parts (e.g. cylinder piston, valve spool) together with the pressure in each control volume as our states.

As the time domain model is concerned with changes in state as time passes, the model consists of differential equations in which the time derivative of all the states occur. The position and velocity of a mechanical mass moving is present when setting up Newton's Second Law:

$$M\ddot{x} + B\dot{x} + kx = \Sigma F. \tag{11.1}$$

The flow continuity equation forms a differential equation including pressure:

$$Q_{\text{in}} - Q_{\text{out}} = \dot{V} + \frac{V}{\beta(p)}\dot{p}. \tag{11.2}$$

All states in the model are required to be in a differential equation of the model. A temperature state therefore requires a differential equation concerning $\frac{dT}{dt}$.

In addition to the differential equations some algebraic equations complete the model. Often algebraic equations are e.g. used to model flows - especially the orifice equation, pipe flow equation and laminar leakage flows.

11.1.2 Reduced Order Model

As mentioned before, the choice of state in a model is defined by the model creator. Even though the choice of state is free it needs to be reasonable relative to the purpose of the model. When analysing or controlling the motion of actuators in a fluid power system, pressure in each control volume together with position and velocity of moving mechanical elements are a sound choice of states. This, however, results in two states per moving mechanical element and one for each modelled control volume pressure which may sum to a significant number, complicating later analysis and model based controller design. Some models may be simplified by reducing the number of states either by excluding some or combing two states into one. An example, given in [1], is the introduction of a load pressure, p_{L}, which is a fictive state defined as the pressure working on the A-side in e.g. a cylinder resulting in the cylinder force:

$$F_{\text{C}} = p_{\text{L}}A_{\text{A}} = p_{\text{A}}A_{\text{A}} - p_{\text{B}}A_{\text{B}}, \tag{11.3}$$

where A_{A} and A_{B} are the working areas in which the chamber pressure p_{A} and p_{B} are working respectively. In some cases it may be useful to exclude the position of the cylinder piston; this can be realised by assuming the cylinder chamber volume doesn't change even though the piston moves. As a result of a non-moving piston chamber, volumes in the cylinder do not change and are as such constant.

11.1.3 Linear Model—Time Domain

Enabling system analysis is a major motivation for developing models of physical system. Many analytical tools are restricted to linear systems, e.g. frequency analysis, root locus plot, eigenvalue evaluation. Therefore, linear time invariant approximations of the constructed non-linear model are required. In this note we shall use first order Taylor approximations and then shift to change variables by subtracting the linearisation point value from the approximation. In this way we get a linear system which, depending on linearisation point, may be either a Linear Time Varying (LTV) or a Linear Time Invariant (LTI) system. The classical tools for frequency and stability analysis request an LTI system, which is why this is our goal.

First Order Taylor Approximation

A short recap on first order Taylor approximation on a function $f(\mathbf{x})$ which is dependent on n variables (the state vector $\mathbf{x}$ has dimension (nx1)) is given here:

$$\tilde{f}(\mathbf{x}) = f(\mathbf{x}_0) + \left.\frac{\partial f(\mathbf{x})}{\partial x_1}\right|_{\mathbf{x}_0} (x_1 - x_{10}) + \left.\frac{\partial f(\mathbf{x})}{\partial x_n}\right|_{\mathbf{x}_0} (x_n - x_{n0}) + H(\mathbf{x}), \quad (11.4)$$

where $\mathbf{x}_0 = [x_{10}, x_{20}, \ldots x_{n0}]$ is the linearisation point leading to the function value $f(\mathbf{x}_0)$ at the linearisation point. Furthermore, $H(\mathbf{x})$ is the sum of all higher order terms which are neglected since they include $(x_i - x_{i0})^2, (x_i - x_{i0})^3 \ldots$ which are small as the change in state from the linearisation point $\mathbf{x}_0$ is assumed small in the linear model. We are as such developing a "small" signal model. The $|_{\mathbf{x}_0}$ indicates that the linearisation point values of the state vector must be inserted into the result of the partial derivative, leading to a constant which is multiplied by the change $(x_i - x_{i0})$. Leaving out the higher order terms and recognising the fraction with the partial derivatives to be constants yield an approximation given as;

$$\tilde{f}(\mathbf{x}) = f(\mathbf{x}_0) + k_1(x_1 - x_{10}) + \cdots + k_n(x_n - x_{n0}). \quad (11.5)$$

This approximation is, though, not a linear function which is why we shall introduce change variables for the states.

Illustration of Change Variable Model

The first order Taylor approximation $\tilde{f}(\mathbf{x})$ is made made into a linear model by introducing change variables for the system's states denoted with Δ such that the change variable Δx_i models the change in function value relative to the linearisation point given as:

$$\tilde{f}(\mathbf{x}) - f(\mathbf{x}_0) = k_1(x_1 - x_{10}) + \cdots + k_n(x_n - x_{n0}), \quad (11.6)$$

$$\Delta f(\Delta \mathbf{x}) = k_1 \Delta x_1 + \cdots + k_n \Delta x_n, \quad (11.7)$$

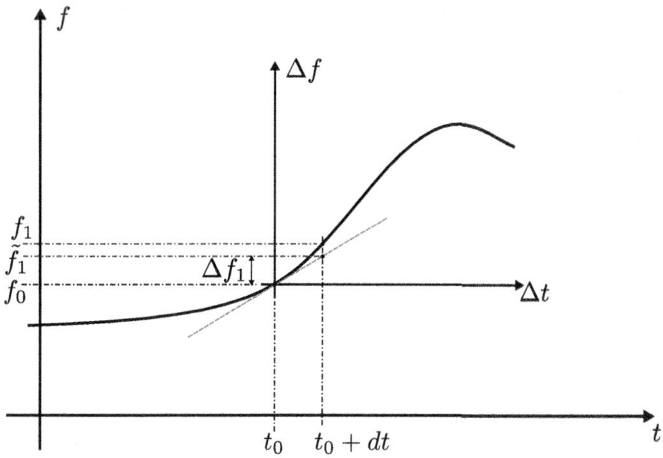

Fig. 11.2 Illustration of linear model in change variables

i.e. the linear approximation of the change in function value is a linear combination of the changes in all of the states.

Figure 11.2 illustrates the relation between a non-linear model value and the linear change variable value for the function value $f(t)$. A linearisation point is chosen at t_0 yielding the function value at the linearisation point of $f_0 = f(t_0)$. The value for the linear model in change variables is at this point $\Delta f(0) = 0$ and $t_0 \Rightarrow \Delta t(t_0) = 0$. In other words, the linear model describing the change variables may be illustrated in a new coordinate system with origin at (t_0, f_0).

Let's compare the values of the two models, the non-linear f and the linear Δf by defining a model error at time $t_1 = t_0 + dt$ as:

$$e_1 = f_1 - (\Delta f_1 + f_0), \tag{11.8}$$

where f_1 is the function value of the non-linear model at t_1 while Δf_1 is the function value at time dt in the change variable model. When we want to compare the two models we need to add the linearisation point value f_0. Generalising for this function of time the error is given as:

$$e = f - \left(\frac{\partial f}{\partial t} dt + f_0\right), \tag{11.9}$$

from which it is evident that the linearisation point must be included when comparing the two models. Furthermore, it is natural that the error will increase when the time dt is increased, especially for highly non-linear functions.

Linearisation Point

The choice of linearisation point is essential for the linear model and the result of the frequency analysis to be useable. When stability analysis is the objective of developing a linear model, the linearisation point is required to be an equilibrium point. This is necessary as linearising at an equilibrium point ensures that the linear model will be a Linear Time Invariant system. If $\mathbf{x}^*$ is an equilibrium point for the system $\dot{\mathbf{x}} = \mathbf{f}(\mathbf{x})$ then the time derivative of the state vector at this point must be zero, which can be expressed as:

$$\mathbf{0} = \dot{\mathbf{x}} = \mathbf{f}(\mathbf{x}^*) \tag{11.10}$$

With an LTI system model one may investigate the stability of the system close to the linearisation point, e.g. one may determine if the equilibrium point is a stable or unstable equilibrium point. The LTI system model may further be used for classical controller design. However, as a non-linear system is linearised the system analysis and controller design is only valid close to the linearisation point. The engineer has to validate the analysis and controller design in the full operation range. As we shall see later the choice of linearisation point greatly influences the conclusions of analysis and the control design process. The choice of linearisation point will be discussed in later sections.

11.2 Motor-Valve Drive

In this section we shall model a fixed displacement motor controlled with a proportional servo valve, i.e. a valve that has internal control of the valve spool position. The supply and tank pressure are assumed constant. Firstly, a time domain model is derived from which a linear model is then constructed.

11.2.1 Time Domain Model

For the motor-valve drive in Fig. 11.3 a simple time domain model consists of one equation of motion, two continuity equations, two orifice equations for the valve flow and a laminar flow equation for the internal motor leakage. Naturally a more complex model may be derived, however, we kept it simple as this is often sufficient for an adequate control design. As mentioned previously, lumped parameter modelling is used and for the current system two control volumes are defined, A and B. They are bounded by the valve and motor port and the hose or pipe walls. From the figure it is seen that supply and tank pressure p_S and p_T respectively are assumed constant. The internal spool position, x_v, control loop is assumed fast relative to the system and as such seen as system input. Later discussion will consider the relevance of using a proper dynamic valve model.

Newton's Second Law for the motor rigidly couple to an inertia load is given as:

$$J\ddot{\theta} = p_A D_m - p_B D_m - B_m\dot{\theta} + \tau_{load}, \tag{11.11}$$

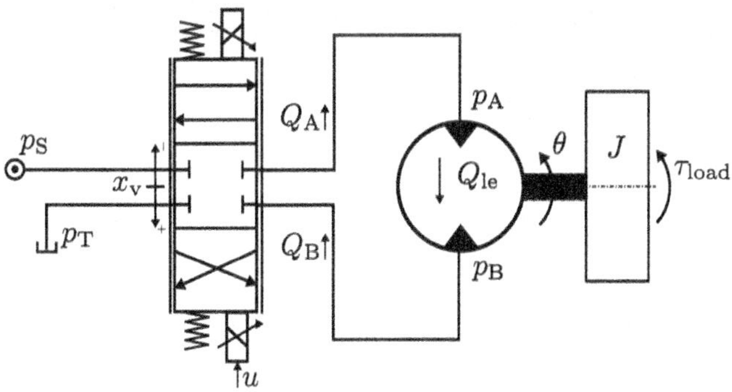

Fig. 11.3 Illustration of a motor-valve drive system, with constant supply pressure

where p_A and p_B is the pressure on each side of the motor, having a displacement of D_m. The total inertia of the motor and load is J while τ_{load} is an external load torque. Friction is modelled as a pure linear viscous damping given by the coefficient B_m multiplied by the rotational velocity of the motor. The pressure gradients in each control volume defined from the valve port to the motor port are modelled with the continuity equation as:

$$\dot{p}_A = \frac{\beta(p_A)}{V_A}\left(Q_A - Q_{le} - D_m\dot{\theta}\right), \tag{11.12}$$

$$\dot{p}_B = \frac{\beta(p_B)}{V_B}\left(Q_{le} - Q_B + D_m\dot{\theta}\right), \tag{11.13}$$

where V_A and V_B is the volume of the two control volumes and $\beta(p)$ is the effective bulk modulus at pressure p. The control volume is mostly comprised of the hose volume. Q_{le} is the internal leakage flow in the motor while Q_A is flow into chamber A through the valve and Q_B is flow out of chamber B through the valve. The orifice equation is used when modelling valve flows and a linear flow equation is used when modelling leakage as the leakage is assumed laminar. The three flows are, then, modelled as:

$$Q_A = x_v \begin{cases} k_{v.PA}\sqrt{|p_S - p_A|}\,\mathrm{sgn}(p_S - p_A) & \text{for } x_v \geq 0 \\ k_{v.AT}\sqrt{|p_A - p_T|}\,\mathrm{sgn}(p_A - p_T) & \text{for } x_v < 0 \end{cases}, \tag{11.14}$$

$$Q_B = x_v \begin{cases} k_{v.BT}\sqrt{|p_B - p_T|}\,\mathrm{sgn}(p_B - p_T) & \text{for } x_v \geq 0 \\ k_{v.PB}\sqrt{|p_S - p_B|}\,\mathrm{sgn}(p_S - p_B) & \text{for } x_v < 0 \end{cases}, \tag{11.15}$$

$$Q_{le} = C_{le}(p_A - p_B), \tag{11.16}$$

where $k_{v.xy}$ is a flow coefficient for the path from x to y, C_{le} is the laminar leakage coefficient and sgn() is the sign function returning -1 or $+1$.

This set of three differential (11.11)–(11.13) and three algebraic flow equations (11.14)–(11.16) constitute the time domain model, where supply and tank pressure are inputs together with valve spool position and load torque.

11.2.2 Linear Model

In this section a linear model is constructed from the non-linear model composed of (11.11)–(11.16). Clearly the orifice equations have two states, either positive or negative valve spool position (zero belongs to positive), i.e. two linear models may be derived, one for positive and one for negative valve spool position. The linear model for positive valve spool position is constructed here and the one for negative is left for the reader to construct using the same approach.

Valve flows Q_A and Q_B are given as:

$$Q_A = x_v k_{v.PA} \sqrt{|p_S - p_A|} \mathrm{sgn}(p_S - p_A) \tag{11.17}$$

$$Q_B = x_v k_{v.BT} \sqrt{|p_B - p_T|} \mathrm{sgn}(p_B - p_T) \tag{11.18}$$

Linearisation by applying first order Taylor approximation of the orifice equations and shifting to change variables as introduced earlier yields:

$$\Delta Q_A = k_{qAx} \Delta x_v + k_{qApS} \Delta p_S - k_{qApS} \Delta p_A, \tag{11.19}$$

$$\Delta Q_B = k_{qBx} \Delta x_v + k_{qBpS} \Delta p_B - k_{qBpB} \Delta p_T, \tag{11.20}$$

where the linearisation constants k_{qAx}, k_{qApS}, k_{qBx}, k_{qBpB} are the partial derivatives evaluated at the linearisation point $\mathbf{x}_0$. The subscript notation for the constants first gives linearised parameter and then the state that the parameter is partially differentiated with respect to such that:

$$k_{qAxv} = \left.\frac{\partial Q_A}{\partial x_v}\right|_{\mathbf{x}_0}, \quad k_{qApS} = \left.\frac{\partial Q_A}{\partial p_S}\right|_{\mathbf{x}_0}. \tag{11.21}$$

Note that k_{qApS} is the constant in front of both Δp_S and Δp_A however with opposite sign. This is chosen to have only positive linearisation constants.

The equation of motion is linear when assuming a fixed displacement motor (D_m is constant) and using only a linear viscous friction model. Change variables are, however, introduced such that it reads:

$$J \Delta\ddot{\theta} = \Delta p_A D_m - \Delta p_B D_m - B_m \Delta\dot{\theta} + \Delta\tau_{load}, \tag{11.22}$$

The A and B volumes are constant as these are the volume of hose connections. If one further assumes the bulk modulus constant (which might be a valid assumption close to $\mathbf{x}_0$), the continuity equations are linear when substituting the linear valve flow in:

$$\Delta\dot{p}_A = \frac{\beta k_{qAx}}{V_A}\Delta x_v + \frac{\beta k_{qAp}}{V_A}\Delta p_S - \frac{\beta k_{qAp}}{V_A}\Delta p_A$$
$$-\frac{\beta C_{le}}{V_A}\Delta p_A + \frac{\beta C_{le}}{V_A}\Delta p_B - \frac{\beta D_m}{V_A}\Delta\dot{\theta}, \tag{11.23}$$
$$\mathrm{d}\dot{p}_B = -\frac{\beta k_{qBx}}{V_B}\Delta x_v - \frac{\beta k_{qBp}}{V_B}\Delta p_B + \frac{\beta k_{qBp}}{V_B}\Delta p_T$$
$$+\frac{\beta C_{le}}{V_B}\Delta p_A - \frac{\beta C_{le}}{V_B}\Delta p_B + \frac{\beta D_m}{V_B}\Delta\dot{\theta}. \tag{11.24}$$

The linear model depends on the following states and inputs: $\mathbf{x} = [\dot{\theta},\ p_A,\ p_A]$, $\mathbf{u} = [x_v,\ p_S,\ p_T,\ \tau_{load}]$. It is important to realise that the model is independent of the motor angular position θ and that this is not a state in the model constructed.

State Space Model

The linear model is set on state space from which we yield the equation:

$$\begin{bmatrix} \dot{p}_A \\ \dot{p}_B \\ \ddot{\theta} \end{bmatrix} = \begin{bmatrix} -\frac{\beta(k_{qApS}+C_{le})}{V_A} & \frac{\beta C_{le}}{V_A} & -\frac{\beta D_m}{V_A} \\ \frac{\beta C_{le}}{V_B} & -\frac{\beta(k_{qBpB}+C_{le})}{V_B} & \frac{\beta D_m}{V_B} \\ \frac{D_m}{J} & \frac{-D_m}{J} & \frac{-B_m}{J} \end{bmatrix} \begin{bmatrix} p_A \\ p_B \\ \dot{\theta} \end{bmatrix}$$
$$+ \begin{bmatrix} \frac{\beta k_{qApS}}{V_A} & 0 & \frac{\beta k_{qAxv}}{V_A} & 0 \\ 0 & \frac{\beta k_{qBpB}}{V_B} & -\frac{\beta k_{qBxv}}{V_B} & 0 \\ 0 & 0 & 0 & \frac{1}{J} \end{bmatrix} \begin{bmatrix} p_S \\ p_T \\ x_v \\ \tau_{load} \end{bmatrix}, \tag{11.25}$$

describing relations between inputs and system states. When searching for a specific output $\mathbf{y}$ one must define the matrices $\mathbf{C}$ and $\mathbf{D}$ in:

$$\mathbf{y} = \mathbf{C}\mathbf{x} + \mathbf{D}\mathbf{u}. \tag{11.26}$$

With inputs defined as $\mathbf{u} = [x_v,\ p_S,\ p_T,\ \tau_{load}]$ one clearly identifies a multiple input system, however when a SISO system is requested some of the inputs may be assumed zero and regarded as disturbances. For the modelled system, setting p_S and p_T to zero is natural as these are assumed constant in the non-linear time domain model and as such zero in change variables. Regarding τ_{load} as a disturbance, the motor velocity could be the output of a SISO system given as:

$$\begin{bmatrix} \dot{p}_{\mathrm{A}} \\ \dot{p}_{\mathrm{B}} \\ \ddot{\theta} \end{bmatrix} = \begin{bmatrix} -\frac{\beta(k_{\mathrm{qApS}}+C_{\mathrm{le}})}{V_{\mathrm{A}}} & \frac{\beta C_{\mathrm{le}}}{V_{\mathrm{A}}} & -\frac{\beta D_{\mathrm{m}}}{V_{\mathrm{A}}} \\ \frac{\beta C_{\mathrm{le}}}{V_{\mathrm{B}}} & -\frac{\beta(k_{\mathrm{qBpB}}+C_{\mathrm{le}})}{V_{\mathrm{B}}} & \frac{\beta D_{\mathrm{m}}}{V_{\mathrm{B}}} \\ \frac{D_{\mathrm{m}}}{J} & \frac{-D_{\mathrm{m}}}{J} & \frac{-B_{\mathrm{m}}}{J} \end{bmatrix} \begin{bmatrix} p_{\mathrm{A}} \\ p_{\mathrm{B}} \\ \dot{\theta} \end{bmatrix} + \begin{bmatrix} \frac{\beta k_{\mathrm{qAxv}}}{V_{\mathrm{A}}} \\ -\frac{\beta k_{\mathrm{qBxv}}}{V_{\mathrm{B}}} \\ 0 \end{bmatrix} x_{\mathrm{v}}, \quad (11.27)$$

$$\dot{\theta} = [0\ 0\ 1] \begin{bmatrix} p_{\mathrm{A}} \\ p_{\mathrm{B}} \\ \dot{\theta} \end{bmatrix}. \quad (11.28)$$

11.3 Cylinder-Valve Drive

In the current section we shall model a standard symmetric cylinder valve drive, for which the supply pressure is assumed controlled to a constant level. The system is seen in Fig. 11.4

It is assumed that the valve is mounted on or close to the cylinder; such connections from valve to cylinder chamber are short and rigid. The connection volumes are hence included in the chamber control volumes when constructing the model.

Firstly, a time domain lumped parameter model is derived, the model is next linearised and presented on state space form. A discussion concerning choice of linearisation point of the cylinder valve drive leads to the following ideas presented [1] and development of a reduced order time domain model. Lastly, a linear model on state space form for the reduced order model is given as compared to the non-reduced model.

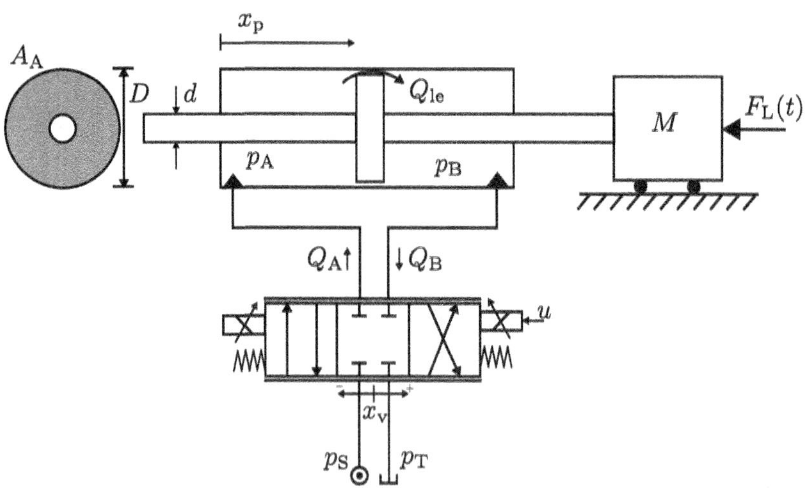

Fig. 11.4 Illustration of cylinder valve drive with constant supply pressure

11.3.1 Time Domain Model

Deriving the time domain model consists, as mentioned, of setting up continuity equations for each control volume, Newton's Second Law for all moving masses and flow equation for fluid flow through various restriction. Recalling the general flow continuity equation, $Q_{\mathrm{in}} - Q_{\mathrm{out}} = \dot{V} + \frac{V}{\beta}\dot{p}$, and using this for chamber A and B each bounded by the servo-valve and the cylinder piston yields:

$$Q_{\mathrm{A}} - Q_{\mathrm{le}} = \dot{x}_{\mathrm{p}} A_{\mathrm{A}} + \frac{V_{\mathrm{Ad}} + x_{\mathrm{p}} A_{\mathrm{A}}}{\beta(p_{\mathrm{A}})} \dot{p}_{\mathrm{A}}, \tag{11.29}$$

$$Q_{\mathrm{le}} - Q_{\mathrm{B}} = -\dot{x}_{\mathrm{p}} A_{\mathrm{A}} + \frac{V_{\mathrm{Bd}} + (L - x_{\mathrm{p}}) A_{\mathrm{A}}}{\beta(p_{\mathrm{B}})} \dot{p}_{\mathrm{B}}, \tag{11.30}$$

where Q_{A} and Q_{B} are flow through the valve into chamber A and flow through the valve out of chamber B respectively, while Q_{le} is the leakage flow from chamber A to B crossing the cylinder piston. External leakage is assumed to be zero. A_{A} is the cylinder piston area, L is the cylinder piston stroke length and x_{p} is cylinder piston position. The dead volumes of chamber A and B are denoted V_{Ad} and V_{Bd} respectively. $\beta(p_{\mathrm{A}})$ and $\beta(p_{\mathrm{B}})$ are the effective bulk modulus of the fluid in chambers A and B respectively.

In describing the motion of the cylinder piston and the rigidly attached load, Newton's Second Law is employed:

$$\ddot{x}_{\mathrm{p}} M = (p_{\mathrm{A}} - p_{\mathrm{B}}) A_{\mathrm{A}} - F_{\mathrm{fr}}(\dot{x}_{\mathrm{p}}) - F_{\mathrm{L}}(t), \tag{11.31}$$

where M is the total mass of the cylinder piston and the load, while $F_{\mathrm{L}}(t)$ is an external load force and $F_{\mathrm{fr}}(\dot{x}_{\mathrm{p}})$ is the friction force. Here a rather simple friction model with coulomb and viscous friction is used:

$$F_{\mathrm{fr}}(\dot{x}_{\mathrm{p}}) = B_{\mathrm{p}} \dot{x}_{\mathrm{p}} + F_{\mathrm{c}} \mathrm{sgn}(\dot{x}_{\mathrm{p}}), \tag{11.32}$$

where B_{p} is the viscous damping coefficient and F_{c} is the scalar value of coulomb friction.

Lastly, equations for fluid flow in the system are set up. Flows through the valve are modelled using the orifice equation while an equation for laminar flows in gaps is used for the leakage flow:

$$\begin{aligned} Q_{\mathrm{A}} &= C_{\mathrm{d}} w x_{\mathrm{v}} \sqrt{\tfrac{2}{\rho} |p_{\mathrm{S}} - p_{\mathrm{A}}|} \mathrm{sgn}(p_{\mathrm{S}} - p_{\mathrm{A}})(x_{\mathrm{v}} \geq 0) \\ &+ C_{\mathrm{d}} w x_{\mathrm{v}} \sqrt{\tfrac{2}{\rho} |p_{\mathrm{A}} - p_{\mathrm{T}}|} \mathrm{sgn}(p_{\mathrm{A}} - p_{\mathrm{T}})(x_{\mathrm{v}} < 0), \end{aligned} \tag{11.33}$$

$$\begin{aligned} Q_{\mathrm{B}} &= C_{\mathrm{d}} w x_{\mathrm{v}} \sqrt{\tfrac{2}{\rho} |p_{\mathrm{B}} - p_{\mathrm{T}}|} \mathrm{sgn}(p_{\mathrm{B}} - p_{\mathrm{T}})(x_{\mathrm{v}} \geq 0) \\ &+ C_{\mathrm{d}} w x_{\mathrm{v}} \sqrt{\tfrac{2}{\rho} |p_{\mathrm{S}} - p_{\mathrm{B}}|} \mathrm{sgn}(p_{\mathrm{S}} - p_{\mathrm{B}})(x_{\mathrm{v}} < 0), \end{aligned} \tag{11.34}$$

$$Q_{\mathrm{le}} = C_{\mathrm{le}}(p_{\mathrm{A}} - p_{\mathrm{B}}). \tag{11.35}$$

A symmetric spool valve with linear opening proportional to the valve spool position, x_v, is used, hence why the flow area in the valve is given as wx_v where w is the opening area gradient. Note that a logic function is used to distinguish between positive and negative valve position (when true $(x_\mathrm{v} \geq 0)$ and $(x_\mathrm{v} < 0)$ gives back 1, while given 0 when false). Note that this is just another implementation of a valve similar to the one used in Sect. 11.2.1. The discharge coefficient, C_d and the oil density, ρ, should be for the specific valve and fluid respectively.

The equations constituting the time domain model are reprinted to give an overview of the model. The model derived consists of: three differential equations describing cylinder piston acceleration and pressure gradients in the two control volume, and three algebraic equations describing flow through the valve and the leakage flow in the cylinder:

$$\ddot{x}_p = \tfrac{1}{M}\left(A_\mathrm{A}(p_\mathrm{A} - p_\mathrm{B}) - B_\mathrm{p}\dot{x}_\mathrm{p} - F_\mathrm{c}\mathrm{sgn}(\dot{x}_\mathrm{p}) - F_\mathrm{L}(t)\right), \tag{11.36}$$

$$\dot{p}_\mathrm{A} = \tfrac{\beta(p_\mathrm{A})}{V_\mathrm{A0}+x_\mathrm{p}A_\mathrm{A}}\left(Q_\mathrm{A} - C_\mathrm{le}(p_\mathrm{A} - p_\mathrm{B}) - \dot{x}_\mathrm{p}A_\mathrm{A}\right), \tag{11.37}$$

$$\dot{p}_\mathrm{B} = \tfrac{\beta(p_\mathrm{B})}{V_\mathrm{B0}-x_\mathrm{p}A_\mathrm{A}}\left(C_\mathrm{le}(p_\mathrm{A} - p_\mathrm{B}) - Q_\mathrm{B} + \dot{x}_\mathrm{p}A_\mathrm{A}\right), \tag{11.38}$$

$$\begin{aligned} Q_\mathrm{A} &= C_\mathrm{d}wx_\mathrm{v}\sqrt{\tfrac{2}{\rho}|p_\mathrm{S} - p_\mathrm{A}|}\mathrm{sgn}(p_\mathrm{S} - p_\mathrm{A})(x_\mathrm{v} \geq 0) \\ &+C_\mathrm{d}wx_\mathrm{v}\sqrt{\tfrac{2}{\rho}|p_\mathrm{A} - p_\mathrm{T}|}\mathrm{sgn}(p_\mathrm{A} - p_\mathrm{T})(x_\mathrm{v} < 0), \end{aligned} \tag{11.39}$$

$$\begin{aligned} Q_\mathrm{B} &= C_\mathrm{d}wx_\mathrm{v}\sqrt{\tfrac{2}{\rho}|p_\mathrm{B} - p_\mathrm{T}|}\mathrm{sgn}(p_\mathrm{B} - p_\mathrm{T})(x_\mathrm{v} \geq 0) \\ &+C_\mathrm{d}wx_\mathrm{v}\sqrt{\tfrac{2}{\rho}|p_\mathrm{S} - p_\mathrm{B}|}\mathrm{sgn}(p_\mathrm{S} - p_\mathrm{B})(x_\mathrm{v} < 0), \end{aligned} \tag{11.40}$$

$$Q_\mathrm{le} = C_\mathrm{le}(p_\mathrm{A} - p_\mathrm{B}). \tag{11.41}$$

This model has four system states: Piston position, Piston velocity and pressures in the two chambers. The model further holds four inputs or disturbances in the form of: valve spool position, supply and tank pressure, and load force.

An important fact to note is the inherent discontinuity in the model. Do you see it? In fact, two might be seen. Of most importance is the discontinuity in the orifice equations introduced by the valve inform of changing the connection path from supply-A to tank-A when the sign of the valve spool position is changed (and likewise for the B side connection). The other discontinuity comes in the modelling of the coulomb friction and is more of a mathematical construction than a physical discontinuity in the friction force.

11.3.2 Linear Model

As the orifice equations have two states, positive or negative valve spool position (zero corresponds to a positive value), two linear models may be derived, one for positive and one for negative valve spool positions. Let's only treat positive valve spool positions (the reader is left to conduct the derivation of the model for negative valve spool positions and comparing the two). Excluding negative valve spool position the orifice equations may be written as:

$$Q_{\mathrm{A}} = C_{\mathrm{d}} w x_{\mathrm{v}} \sqrt{\tfrac{2}{\rho} |p_{\mathrm{S}} - p_{\mathrm{A}}|} \operatorname{sgn}(p_{\mathrm{S}} - p_{\mathrm{A}}), \tag{11.42}$$

$$Q_{\mathrm{B}} = C_{\mathrm{d}} w x_{\mathrm{v}} \sqrt{\tfrac{2}{\rho} |p_{\mathrm{B}} - p_{\mathrm{T}}|} \operatorname{sgn}(p_{\mathrm{B}} - p_{\mathrm{T}}). \tag{11.43}$$

Including the flow equations (orifice and leakage) in the according continuity equations yields that the model now consists of three differential equations given as:

$$\ddot{x}_p = \tfrac{1}{M} \left(A_{\mathrm{A}}(p_{\mathrm{A}} - p_{\mathrm{B}}) - B_{\mathrm{p}} \dot{x}_{\mathrm{p}} - F_{\mathrm{c}} \operatorname{sgn}(\dot{x}_{\mathrm{p}}) - F_{\mathrm{L}}(t) \right), \tag{11.44}$$

$$\begin{aligned} \dot{p}_{\mathrm{A}} = \tfrac{\beta(p_{\mathrm{A}})}{V_{\mathrm{A0}} + x_{\mathrm{p}} A_{\mathrm{A}}} &\left(C_{\mathrm{d}} w x_{\mathrm{v}} \sqrt{\tfrac{2}{\rho} |p_{\mathrm{S}} - p_{\mathrm{A}}|} \operatorname{sgn}(p_{\mathrm{S}} - p_{\mathrm{A}}) \right. \\ &\left. - C_{\mathrm{le}}(p_{\mathrm{A}} - p_{\mathrm{B}}) - \dot{x}_{\mathrm{p}} A_{\mathrm{A}} \right), \end{aligned} \tag{11.45}$$

$$\begin{aligned} \dot{p}_{\mathrm{B}} = \tfrac{\beta(p_{\mathrm{B}})}{V_{\mathrm{B0}} - x_{\mathrm{p}} A_{\mathrm{A}}} &\left(C_{\mathrm{le}}(p_{\mathrm{A}} - p_{\mathrm{B}}) \right. \\ &\left. - C_{\mathrm{d}} w x_{\mathrm{v}} \sqrt{\tfrac{2}{\rho} |p_{\mathrm{B}} - p_{\mathrm{T}}|} \operatorname{sgn}(p_{\mathrm{B}} - p_{\mathrm{T}}) + \dot{x}_{\mathrm{p}} A_{\mathrm{A}} \right). \end{aligned} \tag{11.46}$$

The model is linearised and transformed to a linear model in change variable as discussed in Sect. 11.1.3. The equation of motion is linearised to:

$$M \Delta \ddot{x}_p = \Delta p_{\mathrm{A}} A_{\mathrm{A}} - \Delta p_{\mathrm{B}} A_{\mathrm{A}} - B_{\mathrm{p}} \Delta \dot{x}_{\mathrm{p}} - \Delta F_{\mathrm{L}}(t) \tag{11.47}$$

Let's assume bulk modulus constant and linearise the continuity equations. The model of the continuity equations in the change variables becomes:

$$\begin{aligned} \mathrm{d}\dot{p}_{\mathrm{A}} = & \left.\tfrac{\partial \dot{p}_{\mathrm{A}}}{\partial x_{\mathrm{p}}}\right|_0 \Delta x_{\mathrm{p}} + \left.\tfrac{\partial \dot{p}_{\mathrm{A}}}{\partial x_{\mathrm{v}}}\right|_0 \Delta x_{\mathrm{v}} + \left.\tfrac{\partial \dot{p}_{\mathrm{A}}}{\partial p_{\mathrm{S}}}\right|_0 \Delta p_{\mathrm{S}} \\ & + \left.\tfrac{\partial \dot{p}_{\mathrm{A}}}{\partial p_{\mathrm{A}}}\right|_0 \Delta p_{\mathrm{A}} + \left.\tfrac{\partial \dot{p}_{\mathrm{A}}}{\partial p_{\mathrm{B}}}\right|_0 \Delta p_{\mathrm{B}} + \left.\tfrac{\partial \dot{p}_{\mathrm{A}}}{\partial \dot{x}_{\mathrm{p}}}\right|_0 \Delta \dot{x}_{\mathrm{p}} \end{aligned} \tag{11.48}$$

for the A side chamber and

$$\begin{aligned} \mathrm{d}\dot{p}_{\mathrm{B}} = & \left.\tfrac{\partial \dot{p}_{\mathrm{B}}}{\partial x_{\mathrm{p}}}\right|_0 \Delta x_{\mathrm{p}} + \left.\tfrac{\partial \dot{p}_{\mathrm{B}}}{\partial p_{\mathrm{A}}}\right|_0 \Delta p_{\mathrm{A}} + \left.\tfrac{\partial \dot{p}_{\mathrm{B}}}{\partial p_{\mathrm{B}}}\right|_0 \Delta p_{\mathrm{B}} \\ & + \left.\tfrac{\partial \dot{p}_{\mathrm{B}}}{\partial x_{\mathrm{v}}}\right|_0 \Delta x_{\mathrm{v}} + \left.\tfrac{\partial \dot{p}_{\mathrm{B}}}{\partial p_{\mathrm{T}}}\right|_0 \Delta p_{\mathrm{T}} + \left.\tfrac{\partial \dot{p}_{\mathrm{B}}}{\partial \dot{x}_{\mathrm{p}}}\right|_0 \Delta \dot{x}_{\mathrm{p}} \end{aligned} \tag{11.49}$$

for the B side chamber, where the linearisation constants given as the partial derivatives $\left.\frac{\partial \text{function}}{\partial \text{state}_i}\right|_0$ are evaluated at a given linearisation point and as such become a constant. In deriving the model for positive valve spool positions we yield the following linearisation constants:

$$\left.\frac{\partial \dot{p}_{\text{A}}}{\partial x_{\text{p}}}\right|_0 = \frac{-A_{\text{A}}\beta_0}{(V_{\text{A0}}+x_{\text{p0}}A_{\text{A}})^2}\left(x_{\text{v0}}C_{\text{d}}w\sqrt[\pm]{\frac{2}{\rho}|p_{\text{S0}}-p_{\text{A0}}|}-C_{\text{le}}(p_{\text{A0}}-p_{\text{B0}})-\dot{x}_{\text{p0}}A_{\text{A}}\right)$$
$$= C_{\text{HA0}}k\text{qAxp} \quad (11.50)$$

$$\left.\frac{\partial \dot{p}_{\text{A}}}{\partial x_{\text{v}}}\right|_0 = \frac{\beta_0}{V_{\text{A0}}+x_{\text{p0}}A_{\text{A}}}\left(C_{\text{d}}w\sqrt[\pm]{\frac{2}{\rho}|p_{\text{S0}}-p_{\text{A0}}|}\right) = C_{\text{HA0}}k_{\text{qAxv}} \quad (11.51)$$

$$\left.\frac{\partial \dot{p}_{\text{A}}}{\partial p_{\text{S}}}\right|_0 = \frac{\beta_0}{V_{\text{A0}}+x_{\text{p0}}A_{\text{A}}}\left(\frac{C_{\text{d}}wx_{\text{v0}}\frac{2}{\rho}}{2\sqrt[\pm]{\frac{2}{\rho}|p_{\text{S0}}-p_{\text{A0}}|}}\right) = C_{\text{HA0}}k_{\text{qAp}} \quad (11.52)$$

$$\left.\frac{\partial \dot{p}_{\text{A}}}{\partial p_{\text{A}}}\right|_0 = \frac{\beta_0}{V_{\text{A0}}+x_{\text{p0}}A_{\text{A}}}\left(\frac{-x_{\text{v0}}C_{\text{d}}w\frac{2}{\rho}}{2\sqrt[\pm]{\frac{2}{\rho}|p_{\text{S0}}-p_{\text{A0}}|}}-C_{\text{le}}\right) = C_{\text{HA0}}(-k_{\text{qAp}}-C_{\text{le}}) \quad (11.53)$$

$$\left.\frac{\partial \dot{p}_{\text{A}}}{\partial p_{\text{B}}}\right|_0 = \frac{C_{\text{le}}\beta_0}{V_{\text{A0}}+x_{\text{p0}}A_{\text{A}}} = C_{\text{HA0}}C_{\text{le}} \quad (11.54)$$

$$\left.\frac{\partial \dot{p}_{\text{A}}}{\partial \dot{x}_{\text{p}}}\right|_0 = \frac{-A_{\text{A}}\beta_0}{V_{\text{A0}}+x_{\text{p0}}A_{\text{A}}} = -C_{\text{HA0}}A_{\text{A}} \quad (11.55)$$

for the A side chamber and:

$$\left.\frac{\partial \dot{p}_{\text{B}}}{\partial x_{\text{p}}}\right|_0 = \frac{A_{\text{A}}\beta_{B0}}{(V_{\text{B0}}-x_{\text{p0}}A_{\text{A}})^2}\left(C_{\text{le}}(p_{\text{A0}}-p_{\text{B0}})-x_{\text{v0}}C_{\text{d}}w\sqrt[\pm]{\frac{2}{\rho}|p_{\text{B0}}-p_{\text{T0}}|}+\dot{x}_{\text{p0}}A_{\text{A}}\right)$$
$$= C_{\text{HB0}}k_{\text{qBxp}} \quad (11.56)$$

$$\left.\frac{\partial \dot{p}_{\text{B}}}{\partial p_{\text{A}}}\right|_0 = \frac{C_{\text{le}}\beta_{B0}}{V_{\text{B0}}-x_{\text{p0}}A_{\text{A}}} = C_{\text{HB0}}C_{\text{le}} \quad (11.57)$$

$$\left.\frac{\partial \dot{p}_{\text{B}}}{\partial p_{\text{B}}}\right|_0 = \frac{\beta_{B0}}{V_{\text{B0}}-x_{\text{p0}}A_{\text{A}}}\left(-C_{\text{le}}-\frac{x_{\text{v0}}C_{\text{d}}w\frac{2}{\rho}}{2\sqrt[\pm]{\frac{2}{\rho}|p_{\text{B0}}-p_{\text{T0}}|}}\right) = C_{\text{HB0}}(-C_{\text{le}}-k_{\text{qBp}}) \quad (11.58)$$

$$\left.\frac{\partial \dot{p}_{\text{B}}}{\partial x_{\text{v}}}\right|_0 = \frac{\beta_{B0}}{V_{\text{B0}}-x_{\text{p0}}A_{\text{A}}}\left(-C_{\text{d}}w\sqrt[\pm]{\frac{2}{\rho}|p_{\text{B0}}-p_{\text{T0}}|}\right) = -C_{\text{HB0}}k_{\text{qBxv}} \quad (11.59)$$

$$\left.\frac{\partial \dot{p}_{\text{B}}}{\partial p_{\text{T}}}\right|_0 = \frac{\beta_{B0}}{V_{\text{B0}}-x_{\text{p0}}A_{\text{A}}}\left(-\frac{-x_{\text{v0}}C_{\text{d}}w\frac{2}{\rho}}{2\sqrt[\pm]{\frac{2}{\rho}|p_{\text{B0}}-p_{\text{T0}}|}}\right) = C_{\text{HB0}}k_{\text{qBp}} \quad (11.60)$$

$$\left.\frac{\partial \dot{p}_{\text{B}}}{\partial \dot{x}_{\text{p}}}\right|_0 = \frac{A_{\text{A}}\beta_{B0}}{V_{\text{B0}}-x_{\text{p0}}A_{\text{A}}} = C_{\text{HB0}}A_{\text{A}} \quad (11.61)$$

for the B side chamber, where the hydraulic capacitances of the A and B side chamber in the linearisation point are:

$$C_{\mathrm{HA0}} = \frac{\beta_{\mathrm{A0}}}{V_{\mathrm{A0}} + x_{\mathrm{p0}} A_{\mathrm{A}}}, \tag{11.62}$$

$$C_{\mathrm{HB0}} = \frac{\beta_{B0}}{V_{\mathrm{B0}} - x_{\mathrm{p0}} A_{\mathrm{A}}}, \tag{11.63}$$

respectively. It is left for the reader to substitute the linearisation constant into (11.48) and (11.49) to get a model with linearisation constants. We shall instead present the linear model in change variables in the state space matrix notation.

State Space Model

The linear model in change variables is present in the common state space form:

$$\begin{aligned} \dot{\mathbf{x}} &= \mathbf{Ax} + \mathbf{Bu}, \\ \mathbf{y} &= \mathbf{Cx}, \end{aligned} \tag{11.64}$$

where the state vector and system matrices for the current model of the symmetrical cylinder valve drive are given as:

$$\mathbf{x} = \begin{bmatrix} x_{\mathrm{p}} & \dot{x}_{\mathrm{p}} & p_{\mathrm{A}} & p_{\mathrm{B}} \end{bmatrix}^T \quad , \quad \dot{\mathbf{x}} = \begin{bmatrix} \dot{x}_{\mathrm{p}} & \ddot{x}_{\mathrm{p}} & \dot{p}_{\mathrm{A}} & \dot{p}_{\mathrm{B}} \end{bmatrix}^T \quad , \quad \mathbf{u} = \begin{bmatrix} p_{\mathrm{S}} & p_{\mathrm{T}} & x_{\mathrm{v}} & F_{\mathrm{L}} \end{bmatrix}^T \tag{11.65}$$

$$\mathbf{A} = \begin{bmatrix} 0 & 1 & 0 & 0 \\ 0 & -\frac{B_{\mathrm{p}}}{m} & \frac{A_{\mathrm{A}}}{m} & -\frac{A_{\mathrm{A}}}{m} \\ C_{\mathrm{HA0}} k_{\mathrm{qAxp}} & -C_{\mathrm{HA0}} A_{\mathrm{A}} & -C_{\mathrm{HA0}}(C_{\mathrm{le}} + k_{\mathrm{qAp}}) & C_{\mathrm{HA0}} C_{\mathrm{le}} \\ C_{\mathrm{HB0}} k_{\mathrm{qBxp}} & C_{\mathrm{HB0}} A_{\mathrm{A}} & C_{\mathrm{HB0}} C_{\mathrm{le}} & C_{\mathrm{HB0}}(C_{\mathrm{le}} - k_{\mathrm{qBp}}) \end{bmatrix} \tag{11.66}$$

$$\mathbf{B} = \begin{bmatrix} 0 & 0 & 0 & 0 \\ 0 & 0 & 0 & \frac{1}{m} \\ C_{\mathrm{HA0}} k_{\mathrm{qAp}} & 0 & C_{\mathrm{HA0}} k_{\mathrm{qAxv}} & 0 \\ 0 & C_{\mathrm{HB0}} k_{\mathrm{qBp}} & -C_{\mathrm{HB0}} k_{\mathrm{qBxv}} & 0 \end{bmatrix} \tag{11.67}$$

The output matrix is left out as we may choose various outputs depending on the tasks for which one will use the model.

11.3.3 Reduced Order Time Domain Model—Symmetric Cylinder

Let's discuss the reduced order model for a cylinder valve drive as modelled above. In [1] Merritt shows how a reduced order model for a symmetrical cylinder controlled with a matched and symmetric valve is derived. By defining a load pressure p_{L} and a load flow Q_{L} as:

$$p_L = p_A - p_B, \tag{11.68}$$
$$Q_A = \frac{Q_A + Q_B}{2}, \tag{11.69}$$

and using a steady state assumption in the derivation Merritt derives as model similar to:

$$\dot{p}_L = \frac{4\beta}{V_t}\left(Q_L - \dot{x}_p A_A - Q_{le}\right), \tag{11.70}$$
$$\ddot{x}_p = \frac{1}{M}\left(A_A p_L - B_p \dot{x}_p - F_c \mathrm{sgn}(\dot{x}_p) - F_L(t)\right), \tag{11.71}$$
$$Q_L = C_d w x_v \sqrt{\frac{1}{\rho}\left|p_S - \frac{x_v}{|x_v|}p_L - p_T\right|}\,\mathrm{sgn}\left(p_S - \frac{x_v}{|x_v|}p_L - p_T\right), \tag{11.72}$$

where the total volume V_t is $V_A + V_B$.

State Space Model of Reduced Order System—Symmetric Cylinder

The state space representation of the reduced order linear model is given as:

$$\begin{bmatrix} \dot{p}_L \\ \ddot{x}_p \end{bmatrix} = \begin{bmatrix} -\frac{4\beta(k_{qp}+C_{le})}{V_t} & -\frac{4\beta A_A}{V_t} \\ \frac{A_A}{M} & -\frac{B_p}{M} \end{bmatrix} \begin{bmatrix} p_L \\ \dot{x}_p \end{bmatrix} + \begin{bmatrix} \frac{4\beta k_{qp}}{V_t} & \frac{4\beta k_{qx}}{V_t} & 0 \\ 0 & 0 & \frac{1}{M} \end{bmatrix} \begin{bmatrix} p_S \\ x_v \\ F_L \end{bmatrix}. \tag{11.73}$$

It is significant to notice that the reduced order system model does not depend on the cylinder piston position in the current form. This is partly due to the choice of using the total volume (i.e. linearisation in the centre position) which yields a model where the volume is unchanged with changing piston position. The non-reduced model on the other hand encounters linearisation constants that change with linearisation point (hydraulic capacitance). Furthermore, a position feedback is seen for the non-reduced model, however, the linearisation constants k_{qAxp} and k_{qBxp} may be zero if the linearisation point is chosen properly.

Reduced Order Time Domain Model—Differential Cylinder

Load pressure is defined from the force exerted by the fluid on the cylinder piston such that:

$$F_{cyl} = A_A p_L = A_A p_A - A_B p_B, \tag{11.74}$$
$$p_L = p_A - \frac{A_B}{A_A} p_B. \tag{11.75}$$

With this one can imagine that a reduced order model can naturally be derived for a differential cylinder. This has e.g. been shown in [2], where the hydraulic capacitance in the reduced order model changes with the linearisation point for cylinder piston position. The structure of the model is, however, similar for the symmetric and differential cylinder; they differ though by the linearisation constants.

11.4 Solving the Model

Models describing how a dynamic system behaves are a collection of differential equations. The system being modelled determines which, how many and how the equations are coupled. Furthermore, the detail level chosen by the engineer building the model influences the choice of equation. For a simple system like a Mass-Spring-Damper with an external force input only one second order differential equation is needed to model its movement:

$$M\ddot{x}(t) + B\dot{x}(t) + kx(t) = F(t).$$

Solving the equation means to find $x(t)$ which is the position of the mass as a function of time. With this solution one may say where the mass will be at time t. But how does one find this solution, $x(t)$? For a simple force input and known initial condition ($x(0)$ and $\dot{x}(0)$) one may be able to derive an analytical solution. Suppose the force is constant zero, $F(t) = 0$ and the initial conditions are known e.g. $x(0) = 1$ and $\dot{x}(0) = 0$ the analytical solution may be derived rather "easily" (everything is easy when you know how). Likewise the solution may be "easily" derived with a sinus wave force $F(t) = A\sin(\omega t)$. But, what if the force input becomes much more complex or maybe discrete in nature? What if the damping and spring constant become time dependent e.g. $k = k_1 x$? Even for such a simple system it can become rather complex to derive an analytical solution for $x(t)$. Imagine how to derive the analytical solution for the states in our rather simple Motor-Valve Drive:

$$J\ddot{\theta} = p_A D_m - p_B D_m - B_m\dot{\theta} + \tau_{load},$$

$$\dot{p}_A = \frac{\beta(p_A)}{V_A}\left(Q_A - C_{le}(p_A - p_B) - D_m\dot{\theta}\right),$$

$$\dot{p}_B = \frac{\beta(p_B)}{V_B}\left(C_{le}(p_A - p_B) - Q_B + D_m\dot{\theta}\right),$$

$$Q_A = x_v \begin{cases} k_{v.PA}\sqrt{|p_S - p_A|}\,\mathrm{sgn}(p_S - p_A) & \text{for } x_v \geq 0 \\ k_{v.AT}\sqrt{|p_A - p_T|}\,\mathrm{sgn}(p_A - p_T) & \text{for } x_v < 0 \end{cases},$$

$$Q_B = x_v \begin{cases} k_{v.BT}\sqrt{|p_B - p_T|}\,\mathrm{sgn}(p_B - p_T) & \text{for } x_v \geq 0 \\ k_{v.PB}\sqrt{|p_S - p_B|}\,\mathrm{sgn}(p_S - p_B) & \text{for } x_v < 0 \end{cases}.$$

What is the solution? What are the functions $\dot{\theta}(t)$, $p_A(t)$ and $p_B(t)$ for some given inputs $p_S(t)$ $p_T(t)$, $x_v(t)$, $\tau_{Load}(t)$? Can you derive the analytical solution if we simplify to a constant supply and tank pressure and let the load torque be zero? With the wise words of 'never say never' in mind, let's say it is rather complex to derive an analytical solution, especially if the valve input is complex.

Instead of seeking an analytical solution to the question, what is $x(t)$ for a given input force $F(t)$ and initial condition $x(0)$ and $\dot{x}(0)$? One may simulate the numerical solution true for one set of initial conditions and one input time series. This is done by time integration, often using a tool like MatLab or MatLab SIMULINK to perform the calculations. However, one may easily code a time simulation using a simple integration method like Euler integration.

Time simulation a Mass-Spring-Damper-MatLab Script

Let's construct a simple MatLab Script that will perform a time simulation of the mass spring damper system, with a mass damping coefficient and spring constant of, $M = 1$ kg, $B = 0.7\frac{\text{Ns}}{\text{m}}$ and $k = 10\frac{\text{N}}{\text{m}}$ and the input force $F(t)$. Let's assume initial conditions to be $\dot{x} = 0\frac{\text{m}}{\text{s}}$ and $x = 1$ m. The input force is initially $F = 0$ N and then at time $t = 20$ s it is step to 10 N. The code shown below will simulate this system under such conditions from time 0 to time 40s. The code will return three vectors: a time, velocity and position vector.

```
% Model parameters
  M = 1;        % [kg]
  B = 0.7;      % [Ns/m]
  k = 10;       % [N/m]

% Time Step and simulation time
Ts = 0.01;              % [s]
SimulationTime = 40; % [s]

% Initial conditions
  dx_old = 0; x_old = 1;

% Time Simulation Loop
for n = 1:SimulationTime/Ts

% Force Input
  F = heaviside(n*Ts-20)*10;

% Newton's Second Law
  ddx = 1/M * (F - B*dx_old - k*x_old);

% Integration
  dx(n)  = ddx*Ts + dx_old;
  x(n)  = dx(n)*Ts + x_old;

  dx_old = dx(n);
  x_old = x(n);
end
```

Simulation Results

Time domain simulation gives values for all system states at every time instant in the simulation. The time time domian simulation hence predict how the modelled system will behave under some given conditions, such that one may predict the position and velocity of the mass too as in Fig. 11.5 for this specific input force and these initial conditions. One cannot, however, say anything with certainty about how the modelled system will behave for another input or set of initial conditions. To answer that, one would need to run a new simulation with said conditions.

One may perhaps expect a similar solution for a similar input but one cannot be sure. Later we shall familiarise ourselves with tools like frequency analysis and eigenvalues which we can use to analyse systems for a broader set of conditions.

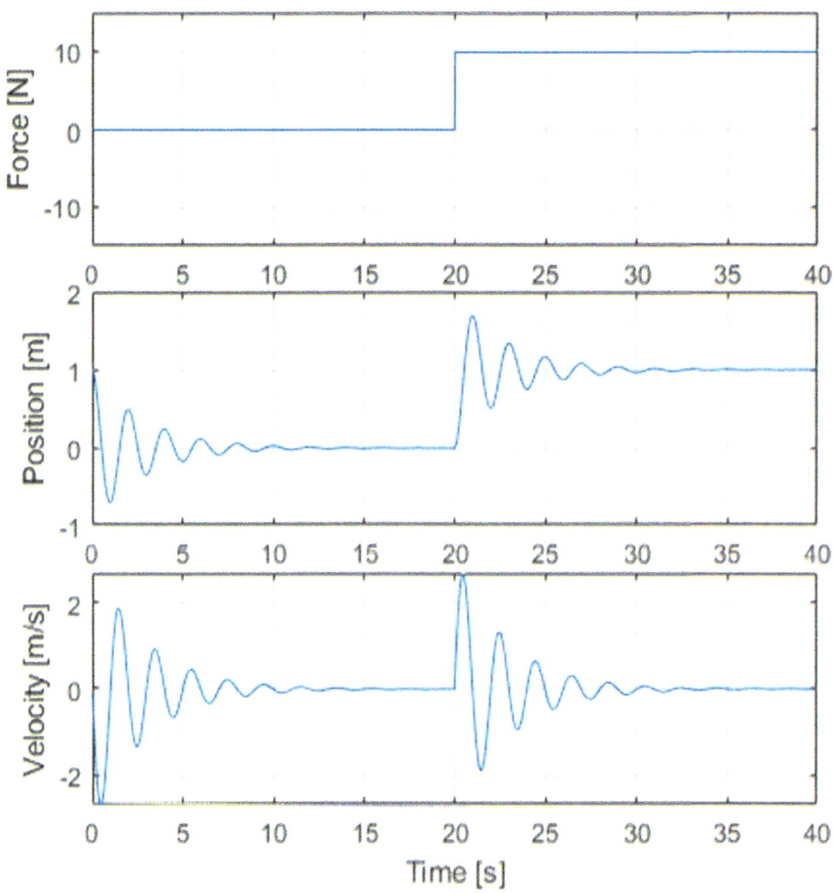

Fig. 11.5 Results from the time domain simulation model. Top to bottom, Input force, mass position and velocity

References

1. Merritt H (1967) Hydraulic control systems
2. Hansen AH, Schmidt L, Pedersen HC, Andersen TO (2016) A generic model based tracking controller for hydraulic valve-cylinder drives. In: Proceedings of the 9th FPNI PhD symposium on fluid power, FPNI2016, Florianopolis-sc, Brazil, 26–28 Oct 2016

Chapter 12
Steady State Analysis

Abstract This chapter introduces a cookbook method for performing a steady state system analysis, followed by two examples. First a simple system easily solvable by hand, and secondly an equally simple system, however non-solvable by hand. The method is highly applicable in a design phase, where the design engineer gets all flows, actuator velocities, pressure levels such that one can calculate power request, power losses and system efficiency.

Steady state analysis is a “tool” used to calculate the steady state value of system states when a constant input is given and the transient response has settled. With the numerical time simulation of a time domain model one gets the behaviour of system states from time zero to the end of simulation. Hence both the initial transient phase and the later steady response if the input is held constant. In steady state analysis only one state value is calculated for each state; the time development of a given state is not given in this analysis.

A steady state analysis may be used while designing a new system to evaluate how the system performs and whether it behaves as intended under a constant input. This tool is limited to examine steady state situations only, meaning that system inputs must be constant and the transient response must have died out. An example could be to calculate the system states in the Motor-Valve Drive when given a constant valve spool position, with constant supply and tank pressure and load torque. Imagine you are setting the valve reference value to $x_v = 0.2$ and then waiting until motor velocity and chamber pressures have settled at constant levels. The state values at this point in time is your solution to a steady state analysis.

We shall investigate steady state analysis by first introducing the algorithm and next using the algorithm on two rather simple systems. As we shall see, the numerical tool comes in handy when solving slightly complex systems and especially when multiple loading conditions or system designs are to be analysed.

A. H. Hansen, *Fluid Power Systems*, Fluid Mechanics and Its Applications 129,
https://doi.org/10.1007/978-3-031-15089-0_12

12.1 The Algorithm

The algorithm for performing a steady state analysis is formed in seven steps:

1. Draw the system diagram.
2. Identify pressure nodes/control volumes.
3. Identify components that bound the pressure nodes/control volumes and define flow directions through these.
4. Choose/guess configurations for all components with more than one mode of operation (Valve State, spool position).
5. Set up equations for the system:
    - Flow continuity for each pressure node.
    - Flow continuity for each pump and actuator.
    - Static force/torque equilibrium for actuators and pumps.
    - Flow equations for flow through restrictions (orifices, filter, hoses etc.).
    - Equations associated with configuration guesses in 4.
    - Static equilibrium for spool and poppet valves.
6. Solve the equations numerically.
7. Evaluate the results: are the computed parameters physically meaningful?

    Yes—Analysis is completed.
    No—Go to point 4 and try out a new configuration set. If all configuration combinations have been tested the analysis is not possible.

Let's now put the 7 step algorithm to work. In the first example the steady state system parameters are calculated together with the process of constructing the equations (this method is used when calculating using pen and paper). Conversely in the second example all steady state system equations are setup and hereafter solve numerically at ones.

12.2 Simple Differential Cylinder System

In the first example a 40 × 25 × 400 mm differential cylinder is considered. It is "controlled" with a 4/3 way directional valve as shown in Fig. 12.1. The directional valve has a discharge coefficient of 0.7 for all paths, the opening areas of the paths in position a are both 18 mm^2. The pump is running at 2000rpm and has a displacement of 25cm^3. The load to be raised weigh 2000kg and the pressure relief valve is set to 250 bar. The oil density is assumed to be 900 $\frac{\text{kg}}{\text{m}^3}$.

In Fig. 12.1 on the left we have the fluid power diagram of the system. On the right we have defined pressure nodes and flow directions. The pressure nodes are bounded by: ① the pump, pressure relief valve and the directional valve; ② the directional valve and the cylinder piston; ③ the cylinder piston and the directional valve; ④ the directional valve, the pressure relief valve and the tank.

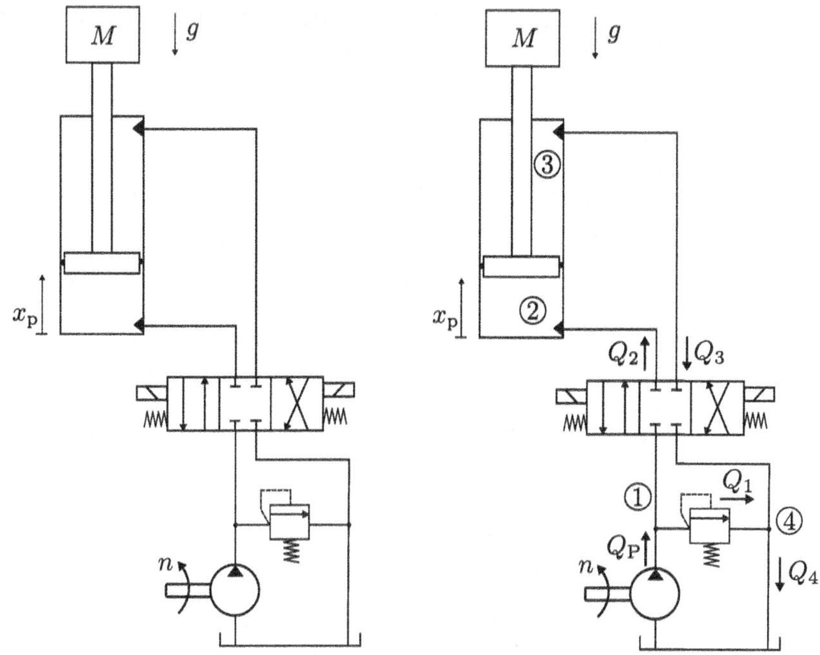

Fig. 12.1 Diagram of example system

In this first example we shall compute the results as we set up the equations. To enable this we need to set up flow equations from the pump through the system to the tank and hereafter the pressure equations from the tank through the system to the pump. However, initially, as when numerical methods are used in step 4, a configuration guess is required. Here we say that the directional valve is in position a (raising the load) and we guess that the pressure relief valve is closed. Further, we assume all components to be ideal ($\eta = 1$).

The continuity equation for pressure node 1 is:

$$Q_P - Q_1 - Q_2 = \dot{V} + \frac{V_1}{\beta}\dot{p}. \tag{12.1}$$

Here we should recall that we are doing a steady state analysis, which is why the transient response is excluded, yielding:

$$Q_P - Q_1 - Q_2 = 0. \tag{12.2}$$

Next we use our guess of zero flow through the pressure relief valve together with a mode for the pump flow

$$Q_1 = 0, \tag{12.3}$$
$$Q_2 = Q_P = Dn. \tag{12.4}$$

Inserting numbers yields:

$$Q_2 = 25\frac{\text{cm}^3}{\text{rev}}\left(\frac{1\text{m}}{100\text{cm}}\right)^3 2000\frac{\text{rev}}{\text{min}}\frac{1\text{min}}{60\text{s}} = 0.000833\frac{\text{m}^3}{\text{s}}, \tag{12.5}$$
$$Q_2 = 50\frac{\text{L}}{\text{min}}. \tag{12.6}$$

The continuity equation for pressure node 2 is:

$$Q_2 - 0 = \dot{V}_2 = \frac{d}{dt}(A_A x_p + V_{A0}), \tag{12.7}$$
$$\dot{x}_p = \frac{Q_2}{A_A} = \frac{0.000833\,\frac{\text{m}^3}{\text{s}}}{(0.040\text{m})^2\frac{\pi}{4}} = 0.6631\frac{\text{m}}{\text{s}}. \tag{12.8}$$

Note that the piston area, A_A, and the initial volume, V_{A0}, are constant. The continuity equation for pressure node 3 is:

$$0 - Q_3 = \dot{V}_3 = \frac{d}{dt}(A_B(L - x_p) + V_{B0}), \tag{12.9}$$
$$Q_3 = A_B\dot{x}_p = \left((0.040\text{mm})^2 - (0.025\text{mm})^2\right)\frac{\pi}{4}\cdot 0.6631\frac{\text{m}}{\text{s}}, \tag{12.10}$$
$$Q_3 = 0.00050781\frac{\text{m}^3}{\text{s}} = 30.47\frac{\text{L}}{\text{min}}. \tag{12.11}$$

The continuity equation for pressure node 4 is, when recalling $Q_1 = 0$:

$$Q_1 + Q_3 - Q_4 = 0, \tag{12.12}$$
$$Q_4 = Q_3 = 30.47\frac{\text{L}}{\text{min}}. \tag{12.13}$$

Now all flows and actuator velocities have been computed. With the flows known, one can compute the pressure drops due to the flows through restrictions in the system, starting from the tank as this is the known pressure in the system. We assume a tank pressure of 1 bar.

When assuming the pipe and hose connections to be loss-free the pressure in node 4 is equal to that of the tank. The pressure loss from node 3 to 4 may be computed using the orifice equation as:

$$Q_3 = C_d A_d\sqrt{\frac{2}{\rho}p_3 - p_4}, \tag{12.14}$$

$$p_3 = \left(\frac{Q_3}{C_d A_d\sqrt{\frac{2}{\rho}}}\right)^2 + p_4, \tag{12.15}$$

$$p_3 = \left(\frac{0.00050781\,\frac{\text{m}^3}{\text{s}}}{0.7\cdot 18\cdot 10{-}6m^2\sqrt{\frac{2}{900\frac{\text{kg}}{\text{m}^3}}}}\right)^2 + 10^5\text{Pa}, \tag{12.16}$$

$$p_3 = 7.31\cdot 10^5\text{Pa} + 10^5\text{Pa} = 8.31\cdot 10^5\text{Pa} = 8.31\text{bar}. \tag{12.17}$$

The pressure in node 2 is computed by use of Newton's Second Law and assuming force balance, hence constant velocity.

$$0 = p_2 A_A - p_3 A_B - Mg, \tag{12.18}$$

$$p_2 = \frac{p_3 A_B + Mg}{A_A}, \tag{12.19}$$

$$p_2 = \frac{8.31 \cdot 10^5 \text{Pa}\big((0.040\text{mm})^2 - (0.025\text{mm})^2\big)\frac{\pi}{4} + 2000\text{kg}9.81\frac{\text{m}}{\text{s}^2}}{(0.040\text{mm})^2\frac{\pi}{4}}, \tag{12.20}$$

$$p_2 = 161.19 \cdot 10^5 \text{Pa} = 161.19\text{bar}. \tag{12.21}$$

Note that an efficiency of 1 is assumed for the cylinder. Next the pressure drop across the directional valve from nodes 1 to 2 is computed utilising the orifice equation once again:

$$Q_2 = C_d A_d \sqrt{\frac{2}{\rho} p_1 - p_2}, \tag{12.22}$$

$$p_1 = \left(\frac{Q_2}{C_d A_d \sqrt{\frac{2}{\rho}}}\right)^2 + p_2, \tag{12.23}$$

$$p_1 = \left(\frac{0.000833\frac{\text{m}^3}{\text{s}}}{0.7 \cdot 18 \cdot 10^-6m^2\sqrt{\frac{2}{900\frac{\text{kg}}{\text{m}^3}}}}\right)^2 + 161.19 \cdot 10^5 \text{Pa}, \tag{12.24}$$

$$p_1 = 19.67 \cdot 10^5 \text{Pa} + 161.19 \cdot 10^5 \text{Pa} = 180.86 \cdot 10^5 \text{Pa} = 180.86\text{bar} \tag{12.25}$$

At this point, pressure in all pressure nodes and the flow between nodes are determined; only the evaluation of the results is lacking.

We can conclude that all flows runs from a high to lower pressure node; we see that $Q_2 > Q_3$ which is expected due to the piston ratio. Finally $p_1 < p_{cr}$, the pump pressure is lower than the crack pressure of the pressure relief valve, hence why the initial guess is valid.

With the pressures and the flows known throughout the system, further analysis may be carried out. The power supplied to the fluid by the pump may be computed and compared with the work performed by the cylinder piston on the mass:

$$\mathcal{P}_{\text{hyd.P}} = Q_P \Delta p_P = D_{\omega P} \Delta p_P \omega, \tag{12.26}$$

$$\mathcal{P}_{\text{hyd.P}} = 0.000833\frac{\text{m}^3}{\text{s}}(180.86 - 1) \cdot 10^5 \text{Pa} = 14.982 \text{ kW}. \tag{12.27}$$

The mechanical power from the cylinder piston to the mass is:

$$\mathcal{P}_{\text{mech}} = Mg\dot{x}_p = 2000 \text{ kg}9.81\frac{m}{s^2}0.66\frac{\text{m}}{\text{s}} = 12.949 \text{ kW} \tag{12.28}$$

This gives an efficiency from fluid power delivered by the pump to mechanical power deliver by the cylinder piston of:

$$\eta = \frac{\mathcal{P}_{\text{mech}}}{\mathcal{P}_{\text{hyd.P}}} = \frac{12.949\ \text{kW}}{14.982\ \text{kW}} = 0.86. \tag{12.29}$$

This is for a system where the piston efficiency is assumed 1 both for the volumetric and the hydro-mechanical efficiency; furthermore the pipes and hoses are assumed loss free. Energy loss is hence solely present in the directional valve where around 2 kW of the power is dissipated as heat. This leads to heating of e.g fluid, fittings and valve blocks.

12.3 Complex Differential Cylinder System

A second and more complex example is given next. Here we shall encounter a system including hose connections of dimensions non-negligible. In the current example hose connections are included between the directional valve and the cylinder chambers. Furthermore, a leakage path is present between the two cylinder chambers, see Fig. 12.2.

The hoses are 10 m long each and have an inner diameter of 15 mm. The leakage gap clearance in the cylinder is measured to 50 μm and the piston thickness is 20 mm. The dynamic viscosity is given as $\mu = 40 \cdot 10^{-6} \cdot 900$ Ns/m. All other parameters are as given in the previous example, hence, the pump is ideal and no friction force is present in the cylinder.

In this example we shall follow the method given in Sect. 12.1. The first three steps brings us to what is given in Fig. 12.2. The fourth step is the guess of configurations; lets assume the directional valve is in position a and raising the piston and that the pressure relief valve is closed.

Step five is to set up all the equations, which we will group in continuity equations, static equilibrium and flow through restriction:

Continuity Equations

The continuity equations for all the pressure nodes are:

$$1: \quad Q_{\text{P}} - Q_1 - Q_2 = 0, \tag{12.30}$$

$$2: \quad Q_2 - Q_3 = 0, \tag{12.31}$$

$$3: \quad Q_3 - Q_{\text{le}} = A_{\text{A}}\dot{x}_{\text{p}}, \tag{12.32}$$

$$4: \quad Q_{\text{le}} - Q_4 = -A_{\text{B}}\dot{x}_{\text{p}}, \tag{12.33}$$

$$5: \quad Q_4 - Q_5 = 0, \tag{12.34}$$

$$6: \quad Q_5 + Q_1 - Q_6 = 0. \tag{12.35}$$

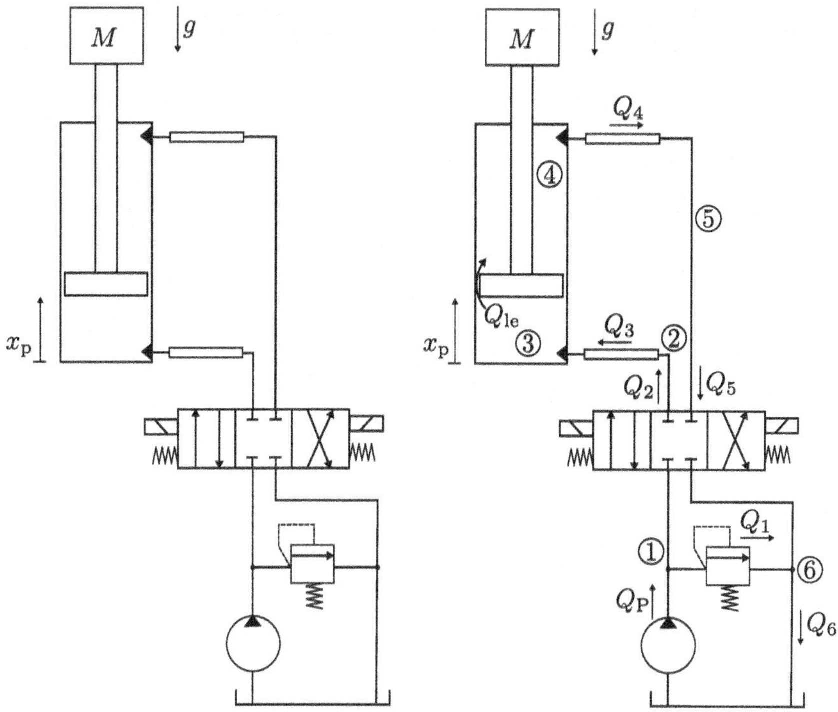

Fig. 12.2 Illustration of case example, including pipes and piston leakage

Static Equilibrium
Applying force balance for the cylinder piston yields:

$$0 = p_3 A_A - p_4 A_B - Mg. \tag{12.36}$$

Flow Through Restrictions
The flow through restrictions as orifices, hoses and gaps are:

$$Q_1 = 0 \tag{12.37}$$
$$Q_2 = C_d A_d \sqrt{\tfrac{2}{\rho}|p_1 - p_2|}\,\text{sgn}(p_1 - p_2) \tag{12.38}$$
$$Q_3 = \frac{\pi d_h^4}{128\mu L_h}(p_2 - p_3) \tag{12.39}$$
$$Q_{le} = C_{le}(p_3 - p_4) \tag{12.40}$$
$$Q_4 = \frac{\pi d_h^4}{128\mu L_h}(p_4 - p_5) \tag{12.41}$$
$$Q_5 = C_d A_d \sqrt{\tfrac{2}{\rho}|p_5 - p_6|}\,\text{sgn}(p_5 - p_6) \tag{12.42}$$

Solving the Equations

Step 6 is to solve the equations. This is done numerically. A system parameter vector $\underline{q}$ is defined such that it consists of all the unknown parameters in the system. All the equations are also combined in a vector Φ such that it can be set equal to zero.

$$\underline{\Phi(\underline{q})} = \underline{0} \tag{12.43}$$

$$\underline{q} = [Q_1,\ Q_2,\ Q_3,\ Q_4, Q_5,\ Q_6,\ p_1,\ p_2,\ p_3,\ p_4,\ p_5,\ p_6,\ \dot{x}_p,\ Q_{le}] \tag{12.44}$$

Solving the steady state equations numerically yields the results shown in table 12.1. With both the equations and the numerical solver at hand, the design engineer may easily review redesigns or other load cases. If the hose diameter is changed to 8 mm the results will be as in table 12.2.

What if the system is overloaded? Let's in the current case increase the weight to 3000 kg. Solving the problem without changing the initial guess of the pressure relief valve state ($Q_1 = 0$) gives the results in Table 12.3. Evaluating these results it is seen that the pressure p_1 just after the pump is 262.66 bar and as this is higher than the

Table 12.1 Results of the example including hose connections and leakage in the piston

Pressure [bar]	Flow$\left[\frac{\text{L}}{\text{min}}\right]$
$p_1 = 184.50$	$Q_1 = 0$
$p_2 = 164.78$	$Q_2 = 50$
$p_3 = 162.36$	$Q_3 = 50$
$p_4 = 10.21$	$Q_4 = 31.28$
$p_5 = 8.70$	$Q_5 = 31.28$
$p_6 = 1.00$	$Q_6 = 31.28$
	$Q_{le} = 2.075$
$\dot{x}_p = 0.636\frac{\text{m}}{\text{s}}$	

Table 12.2 Results of the example including hose connections and leakage in the piston. $d_h = 8$ mm

Pressure [bar]	Flow$\left[\frac{\text{L}}{\text{min}}\right]$
$p_1 = 222.31$	$Q_1 = 0$
$p_2 = 202.63$	$Q_2 = 50$
$p_3 = 172.79$	$Q_3 = 50$
$p_4 = 27.33$	$Q_4 = 31.24$
$p_5 = 8.69$	$Q_5 = 31.24$
$p_6 = 1.00$	$Q_6 = 31.24$
	$Q_{le} = 1.98$
$\dot{x}_p = 0.637\frac{\text{m}}{\text{s}}$	

Table 12.3 Results of the example including hose connections and leakage in the piston, with a load increased to 3000 kg. $d_h = 15$ mm

Pressure [bar]	Flow$\left[\frac{L}{min}\right]$
$p_1 = 262.66$	$Q_1 = 0$
$p_2 = 242.97$	$Q_2 = 50$
$p_3 = 240.59$	$Q_3 = 50$
$p_4 = 10.44$	$Q_4 = 31.69$
$p_5 = 8.91$	$Q_5 = 31.69$
$p_6 = 1.00$	$Q_6 = 31.69$
	$Q_{le} = 3.14$
$\dot{x}_p = 0.62 \frac{m}{s}$	$x_{v,r} = 0.0$mm

Table 12.4 Results of the example including hose connections and leakage in the piston, with a load increased to 3000 kg. $d_h = 15$ mm

Pressure [bar]	Flow$\left[\frac{L}{min}\right]$
$p_1 = 253.80$	$Q_1 = 9.35$
$p_2 = 240.78$	$Q_2 = 40.66$
$p_3 = 238.82$	$Q_3 = 40.66$
$p_4 = 7.58$	$Q_4 = 26.01$
$p_5 = 6.33$	$Q_5 = 26.01$
$p_6 = 1.00$	$Q_6 = 35.35$
	$Q_{le} = 3.15$
$\dot{x}_p = 0.498 \frac{m}{s}$	$x_{v,r} = 0.298$mm

cracking pressure of the relief valve, one must disregard the results as wrong since the initial guess is false. To get correct results one needs to change the assumption $Q_1 = 0$ to:

$$Q_1 = C_d \pi x_x \sqrt{\frac{2}{\rho}|p_1 - p_6|} \text{sgn}(p_1 - p_6), \tag{12.45}$$

and further set a force equilibrium for the relief valve as:

$$0 = (x_{v0} + x_v) k_s - p_1 A_v. \tag{12.46}$$

With this new guess and equations in the model, the results read as in Table 12.4. It is seen that a flow Q_1 is running through the relief valve, i.e. the velocity is significantly lower not only due to leakage flow but also due to the flow through the relief valve. The pressure p_1 after the pump is seen to be higher than the crack pressure; this is due to the non-ideal pressure relief valve model where the pressure difference $p_1 - p_{cr}$ is opening the relief valve against a spring.

With the information given from the steady state analysis the design engineer has great knowledge of system performance. In the sections on actuator and pump modelling it was shown how efficiency may be modelled. These models may and should be easily included in the steady state analysis. We will include this in the exercises on steady state system analysis.

Chapter 13
Frequency Analysis

Abstract This chapter introduces the frequency analysis tool and its application in fluid power systems. The motor-valve drive and the cylinder-valve drive are both linearized, Laplace transformed and put on transfer function form. It is illustrated how the frequency response shown in a bode diagram can be used for parameter studies in fluid power systems, the influence of a change in, e.g., load mass, oil bulk modulus or linearization point are investigated for the two "case" system.

With a lumped parameter simulation model the fluid power design engineer may simulate the behaviour of a given system in the time domain. However, results are only given for the simulated situation. With frequency analysis the design engineer may predict the behaviour of the system for a wide range of inputs, furthermore system stability may be addressed using simple classical control theory.

Frequency analysis is a tool feasible for component, system and controller design. With frequency analysis the engineer may examine the dynamic behaviour of a component or system. A frequency analysis may be applied to components as common as pressure relief valves, where a frequency analysis may predict if a relief valve has the required dynamic behaviour for a given system design. Model based frequency analysis takes offset in a time domain system model, which is linearised and Laplace transformed such that a transfer function from input to output may be constructed, i.e. a linear model is required. With the transfer function at hand numerous analytical tools become applicable.

If no system model exists, frequency analysis may be applied to analyse the system, e.g. by constructing a bode diagram based on the physical system the engineer may yield an experimentally derived transfer function. Various system identification methods may likewise be employed to develop a model.

A brief introduction to the few analysis methods employed in this note is followed by two examples of frequency analysis for the motor-valve and cylinder-valve drive systems modelled in Sects. 11.2 and 11.3 respectively.

A. H. Hansen, *Fluid Power Systems*, Fluid Mechanics and Its Applications 129,
https://doi.org/10.1007/978-3-031-15089-0_13

13.1 Analysis Methods

In performing frequency analysis, we based our studies on linear models which are Laplace transformed, and as such bringing the model into the Laplace domain. The linear model now in the Laplace domain is rearranged such that the requested output is given based on the inputs. Let's see the rod map for the Mass-Spring-Damper system with an external force input for which a the time domain model is given as:

$$M\ddot{x}(t) + B\dot{x}(t) + kx(t) = F(t), \tag{13.1}$$

which, if the coefficients are constants, is already linear, such that a linear model in change variables is:

$$M\Delta\ddot{x}(t) + B\Delta\dot{x}(t) + k\Delta x(t) = \Delta F(t). \tag{13.2}$$

Laplace transforming yields:

$$M(X(s)s^2 - sx(0) - \dot{x}(0)) + B(X(s)s - x(0)) + kX(s) = F(s), \tag{13.3}$$

and if one assumes initial conditions to be zero, $\dot{x}(0) = x(0) = 0$, the Laplace transformed model reads:

$$X(s)(Ms^2 + Bs + k) = F(s). \tag{13.4}$$

A transfer function gives the relation between an output and an input; for models with multiple inputs one must assume those not treated in the transfer function to be zero. The Transfer function as such only gives an output based on this single input; super positioning may be used if a full output is requested. For the Mass-Spring-Damper example only one input exits and the transfer function from force input to mass position as output is therefore given by:

$$X(s) = \frac{1}{Ms^2 + Bs + k}F(s), \tag{13.5}$$

$$\frac{X(s)}{F(s)} = \frac{1}{Ms^2 + Bs + k}. \tag{13.6}$$

Bode Diagram

In performing frequency analysis of systems, bode diagrams are an essential tool that give the frequency dependent system gain and phase shift between input and output. Bode diagrams are not reserved solely for linear systems, however to create a bode diagram based the non-linear time domain model one request several simulation runs of the simulation model or one may "measure" a frequency response on a physical system.

Bode diagrams are based on "measurement" of the output when a system is given a sinus wave as input. The basis is that if a system is given a sinus wave input the output will be a sinus wave, with the frequency, however, of another amplitude and with a phase shift relative to the input. Such that if $r(t) = A\sin(\omega t)$ is the input sinus wave with amplitude A and frequency ω, $y(t) = \hat{A}\sin(\omega t + \phi)$ will be the output sinus wave with amplitude $\hat{A}$, frequency ω and phase shift ϕ. The bode diagram plots the Gain and Phase shift given as:

$$\text{Gain} = \frac{\hat{A}}{A}, \tag{13.7}$$

$$\text{Phase} = \phi. \tag{13.8}$$

If an engineer requests a frequency response of a system one may impose a sinus wave to the system input with various frequencies and compare the input and output to get the gain and phase shift at each frequency. Note, however, that if performed on a non-linear system the input amplitude may highly affect the frequency response.

For a linear system in the Laplace domain, a bode diagram may be constructed by substituting s with $j\omega$, hence the complex operator and input frequency. The Gain is such that the absolute value of the transfer function value at $j\omega$ while the angle gives phase shift. Values for the bode diagram are as such given as:

$$\text{Gain} = \left|\frac{Y(j\omega)}{R(j\omega)}\right|, \tag{13.9}$$

$$\text{Phase} = \angle\frac{Y(j\omega)}{R(j\omega)}. \tag{13.10}$$

Throughout the note and various exercises and examples we shall use the MatLab command $bode(G(s))$ which returns a Bode diagram for the system given by the transfer function $G(s)$.

13.2 Motor-Valve Drive

This section serves as an example of how a frequency analysis may be conducted for a system. The Motor-Valve Drive system of Sect. 11.2 is used and models derived in that section form the basis for the analysis.

Initially, the linear model is Laplace transform to allow for a transfer function which is used for developing frequency responses. Lastly, a study on the influence of parameter variations is performed and discussed.

13.2.1 Transfer Function

Let's now transform the model from Sect. 11.2.2 to the Laplace domain; during transformation we assume initial conditions to be zero:

$$P_A(s)\left(s\frac{V_A}{\beta}+(k_{qAp}+C_{le})\right)=k_{qAx}X_v(s)+k_{qAp}P_S(s)+C_{le}P_B(s)-D_m s\theta(s), \quad (13.11)$$

$$P_B(s)\left(s\frac{V_B}{\beta}+(k_{qBp}+C_{le})\right)=-k_{qBx}X_v(s)+k_{qBp}P_T(s)+C_{le}P_A(s)+D_m s\theta(s), \quad (13.12)$$

$$s\theta(s)\,(Js+B_m)=P_A(s)D_m-P_B(s)D_m+\tau_{load}(s). \quad (13.13)$$

Let's for simplicity assume supply and tank pressure to be constant, where $P_S(s)=P_T(s)=0$. Furthermore let's assume the leakage coefficient to be zero, such that the simplified continuity equations reads:

$$P_A(s)=\frac{k_{qAx}X_v(s)}{s\frac{V_A}{\beta}+k_{qAp}}-\frac{D_m s\theta(s)}{s\frac{V_A}{\beta}+k_{qAp}}, \quad (13.14)$$

$$P_B(s)=-\frac{k_{qBx}X_v(s)}{s\frac{V_B}{\beta}+k_{qBp}}+\frac{D_m s\theta(s)}{s\frac{V_B}{\beta}+k_{qBp}}. \quad (13.15)$$

Combining the two continuity equations (13.14), (13.15) and the equation of motion, (13.13) yields:

$$s\theta(s)=\frac{\left(\frac{D_m k_{qAx}}{s\frac{V_A}{\beta}+k_{qAp}}+\frac{D_m k_{qBx}}{s\frac{V_B}{\beta}+k_{qBp}}\right)}{Js+B_m+\frac{D_m^2}{s\frac{V_A}{\beta}+k_{qAp}}+\frac{D_m^2}{s\frac{V_B}{\beta}+k_{qBp}}}X_v(s)+\frac{1}{Js+B_m+\frac{D_m^2}{s\frac{V_A}{\beta}+k_{qAp}}+\frac{D_m^2}{s\frac{V_B}{\beta}+k_{qBp}}}\tau_{load}(s), \quad (13.16)$$

where the motor velocity $s\theta(s)$ is a function of the valve position $X_v(s)$ and the load torque $\tau_{load}(s)$.

In forming a transfer function from input to output we assume the other input to be zero. When assuming the change in load torque to be zero the transfer function from valve spool position to motor velocity is given as:

$$\frac{s\theta(s)}{X_v(s)}=\frac{\left(\frac{D_m k_{qAx}}{s\frac{V_A}{\beta}+k_{qAp}}+\frac{D_m k_{qBx}}{s\frac{V_B}{\beta}+k_{qBp}}\right)}{Js+B_m+\frac{D_m^2}{s\frac{V_A}{\beta}+k_{qAp}}+\frac{D_m^2}{s\frac{V_B}{\beta}+k_{qBp}}}. \quad (13.17)$$

13.2.2 Frequency Response

The transfer function enables various analysis tools, e.g. Bode plot, root locus. Let's investigate the frequency response of the motor-valve drive using the Bode diagram

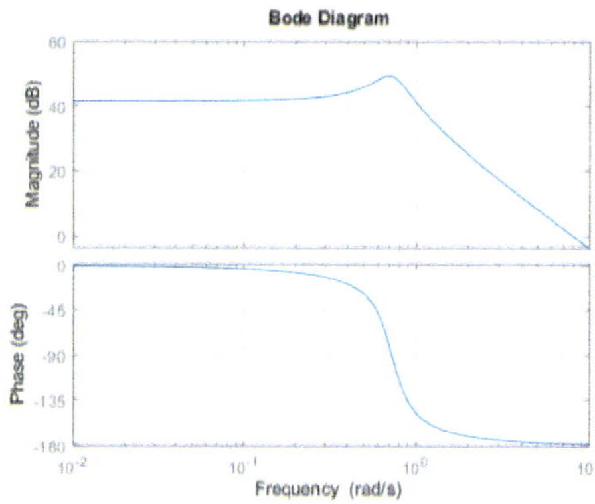

(a) Frequency response for the Motor-Valve Drive

Symbol	Value	Unit
J	10	kgm^2
B_m	1	Nms/rad
V_A	0.001	m^3
V_B	0.001	m^3
D_m	10	cm^3/rev
Q_{nom}	10	L/min
p_{nom}	35	bar
β	10000	bar
p_S	100	bar
p_{nom}	0	bar

(b) System parameter values.

Fig. 13.1 Frequency responses the cylinder valve drive

which plots the system gain and phase shift at varying frequency. Figure 13.1a shows the Bode diagram of (13.17) using the constants given in Fig. 13.1b.

The frequency response indicates an under dampened second order system (smooth phase shift of −180, resonance peak, constant DC gain). One may read of an approximate eigenfrequency of 0.7 radians per second. Let's investigate how variation in some system parameters will affect the frequency response and such system behaviour.

Inertia

Firstly, variations in inertia of the rotating body are varied. A significant effect is seen on the frequency response, see Fig. 13.2, when varying inertia as 2, 4, 6, 8, and 10 kgm^2. An increase in inertia clearly results in small system eigenfrequency while no significant change in either damping nor DC-gain is seen.

Bulk Modulus

Bulk modulus for the utilised fluid depends on e.g. air content and pressure level, such that the same system may experience various linearisation point values for the bulk modulus. Changing the value without regarding a possible cause five frequency responses are given in Fig. 13.3 with increasing bulk modulus from line 1 to 5. Neither damping nor DC-gain is seen affected while the system eigenfrequency on the other hand is largely affected with increasing frequency for increasing bulk modulus.

Valve Spool Position—Linearisation Point

As a last parameter, the valve spool position is varied at the linearisation point; this relates to varying the constant velocity at the linearisation point. Again five responses are given and show an effect, see Fig. 13.4.

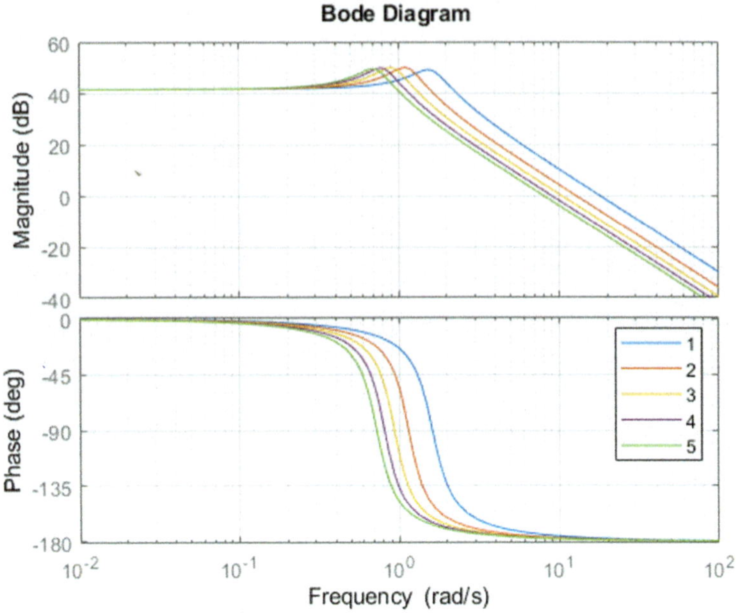

Fig. 13.2 Variation in frequency response for various inertia loads. With $J = [2,\ 4,\ 6,\ 8,\ 10]$ kgm^2 referring to lines [1–5]

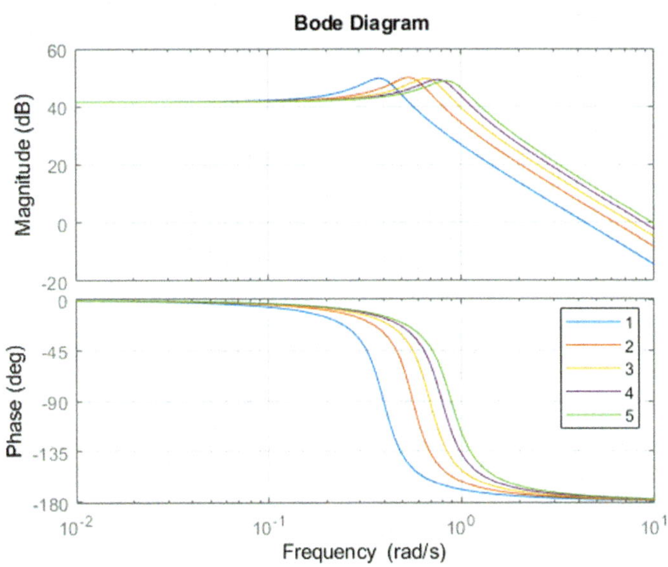

Fig. 13.3 Variation in frequency response for changing bulk modulus for the oil. With $\beta = [3000,\ 6000,\ 9000,\ 12000,\ 15000]$ bar referring to lines [1–5]

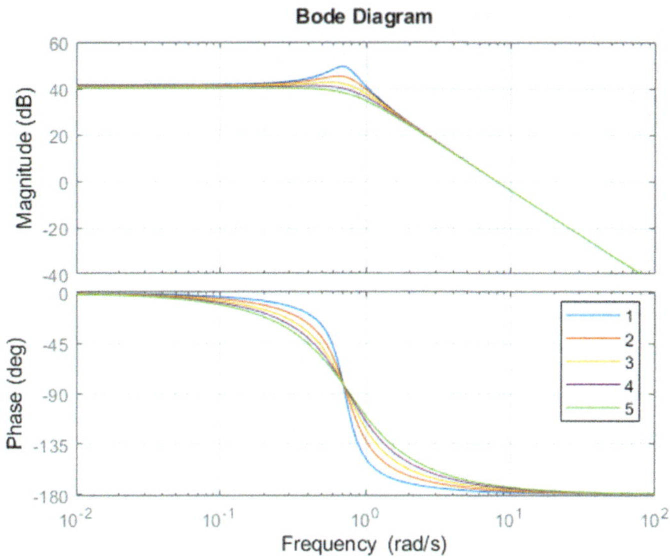

Fig. 13.4 Variation in frequency response for changing valve spool linearisation point. With $x_{v0} = [0.01,\ 0.02,\ 0.03,\ 0.04,\ 0.05]$ referring to lines [1–5]

Variations in valve spool position, i.e. valve opening area, significantly affects damping of the system while close to no effects are seen on the DC-gain and eigenfrequency. A larger valve opening induces more damping.

13.2.3 System Understanding—Motor-Valve Drive

The frequency analysis clearly indicates that the Motor-valve drive system is an "under dampened" second order system. It is, though, evident that one may experience a dampened second order system if one chooses a large valve opening area corresponding to a large motor velocity as linearisation point. It is therefore evident that choice of linearisation point significantly influences controller design basis, as one often develops controllers based on a frequency response. But which response should you base your design on?

The system eigenfrequency is seen to change with the inertia load and bulk modulus. It may therefore be worth considering the possible variations in the current system. Do the inertia, load torque or supply pressure vary under operation?

13.3 Cylinder-Valve Drive

This section serves as an example of how a frequency analysis may be conducted for a system. The Cylinder-Valve Drive system of Sect. 11.3 is used and models derived in that section form the basis for the analysis.

Initially the linear model is Laplace transform to allow for constructing transfer functions. As multiple inputs may be defined, three transfer functions are constructed. The analysis is continued for the transfer function from valve spool position to cylinder piston velocity as this is essential for controller development. Lastly, a study on the influence of parameter variations is performed and discussed.

13.3.1 Transfer Function for the Reduced Order Model

Having linearised the non-linear model, a frequency analysis may be conducted. Firstly, the linear model in change variables is brought to the Laplace domain by Laplace transformation. Secondly, a transfer function describing the relationship between a state and the input is formed. Laplace transforming the reduced order linear model:

$$\mathcal{L}\{\Delta\ddot{x}_{\mathrm{p}}\} = \mathcal{L}\{\tfrac{1}{M}\left(A_{\mathrm{A}}\Delta p_{\mathrm{L}} - B_{\mathrm{p}}\Delta\dot{x}_{p} - \Delta F_{\mathrm{L}}\right)\}, \tag{13.18}$$

$$\mathcal{L}\{\Delta\dot{p}_{\mathrm{L}}\} = \mathcal{L}\left\{\tfrac{4\beta}{V_{\mathrm{t}}}\left(k_{\mathrm{qx}}\Delta x_{\mathrm{v}} + k_{\mathrm{qp}}\Delta p_{\mathrm{S}} - k_{\mathrm{qp}}\Delta p_{\mathrm{L}} - \Delta\dot{x}_{\mathrm{p}}A_{\mathrm{A}} - \Delta p_{\mathrm{L}}C_{\mathrm{le}}\right)\right\}, \tag{13.19}$$

while assuming the initial conditions to be zero, $\Delta\dot{x}_{\mathrm{p}}(0) = \Delta x_{\mathrm{p}}(0) = \Delta p_{\mathrm{L}}(0) = 0$ the model in the Laplace domain may be written as:

$$s^2 X_{\mathrm{p}}(s) = \tfrac{1}{M}\left(A_{\mathrm{A}}P_{\mathrm{L}}(s) - B_{\mathrm{p}}sX_{\mathrm{p}}(s) - F_{\mathrm{L}}(s)\right) \tag{13.20}$$

$$sP_{\mathrm{L}}(s) = \tfrac{4\beta}{V_{\mathrm{t}}}\left(k_{\mathrm{qx}}X_{\mathrm{v}}(s) - k_{\mathrm{qp}}P_{\mathrm{L}}(s) + k_{\mathrm{qp}}P_{\mathrm{S}}(s) - C_{\mathrm{le}}P_{\mathrm{L}}(s) - X_{\mathrm{p}}(s)sA_{\mathrm{A}}\right) \tag{13.21}$$

To enhance an overview of the system the Eqs. (13.20) and (13.21) may be arranged in a block diagram as seen in Fig. 13.5.

Rearranging and reducing the equations gives:

$$X_{\mathrm{p}}(s)s\left(Ms + B_{\mathrm{p}}\right) = A_{\mathrm{A}}P_{\mathrm{L}}(s) - F_{\mathrm{L}}(s), \tag{13.22}$$

$$P_{\mathrm{L}}(s)\left(\tfrac{V_{\mathrm{t}}}{4\beta}s + k_{\mathrm{qp}} + C_{\mathrm{le}}\right) = k_{\mathrm{qx}}X_{\mathrm{v}}(s) + k_{\mathrm{qp}}P_{\mathrm{S}}(s) - X_{\mathrm{p}}(s)sA_{\mathrm{A}}. \tag{13.23}$$

The block diagram following Eqs. (13.22) and (13.23) is seen in Fig. 13.6.

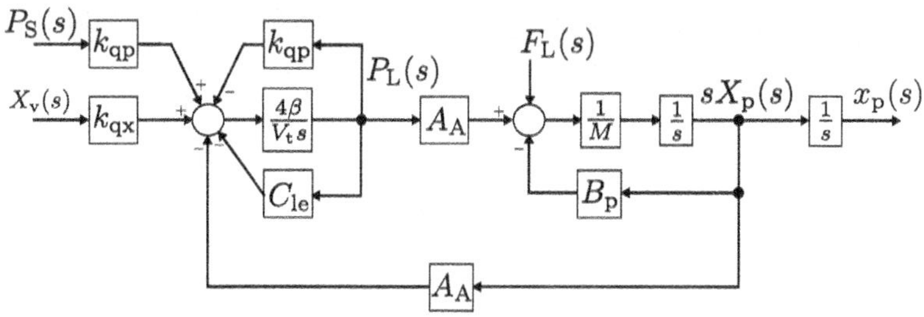

Fig. 13.5 Block diagram for the linear model in the Laplace domain. From (13.20) and (13.21)

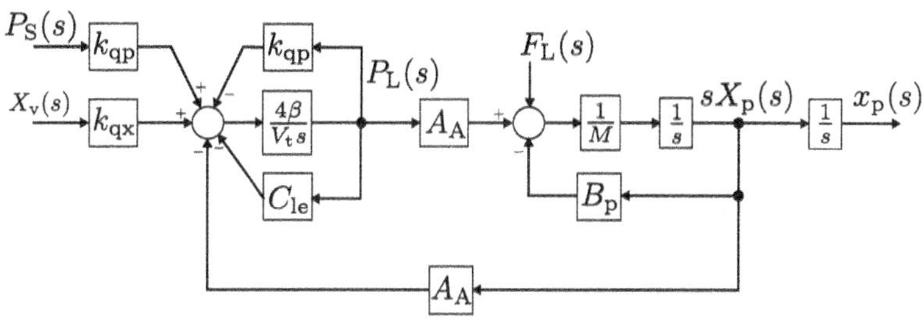

Fig. 13.6 Block diagram for the linear model in the Laplace domain. From (13.22) and (13.23)

Let's derive a transfer function with valve spool position as input and piston velocity as output. Combining Eqs. (13.22) and (13.23) and solving for cylinder piston velocity, $sX_p(s)$ yields:

$$P_L(s) = \frac{k_{qx}X_v(s)+k_{qp}P_S(s)-X_p(s)sA_A}{\frac{V_t}{4\beta}s+k_{qp}+C_{le}}, \tag{13.24}$$

$$sX_p(s) = \frac{A_A P_L(s)-F_L(s)}{Ms+B_p}, \tag{13.25}$$

$$sX_p(s) = \frac{A_A(k_{qx}X_v(s)+k_{qp}P_S(s))-F_L(s)(\frac{V_t}{4\beta}s+k_{qp}+C_{le})}{\frac{MV_t}{4\beta}s^2+\left((k_{qp}+C_{le})M+\frac{B_pV_t}{4\beta}\right)s+B_p(k_{qp}+C_{le})+A_A^2}. \tag{13.26}$$

It can be seen see that cylinder piston velocity $sX_p(s)$ depends on three varying values: valve spool position, supply pressure and load force. However, with a transfer function only one input is permitted. Let's assume that the supply pressure is constant as indicated in Fig. 11.4 and let the load force be seen as a disturbance which is neglected (one may later investigate the influence of supply pressure and load force). Setting $F_L(s) = 0,\ P_S(s) = 0$ (recall that $P_S(s)$ changes from the linearisation point) yields a transfer function from valve spool position to cylinder piston velocity given as:

$$sX_p(s) = \frac{A_A k_{qx}X_v(s)}{\frac{MV_t}{4\beta}s^2+\left((k_{qp}+C_{le})M+\frac{B_pV_t}{4\beta}\right)s+B_p(k_{qp}+C_{le})+A_A^2}. \tag{13.27}$$

One will evidently see that the velocity system is a second order system which on the standard form is:

$$\frac{sX_p(s)}{X_v(s)} = \frac{k_{qx}\frac{4\beta A_A}{MV_t}}{s^2+\left(\frac{4\beta(k_{qp}+C_{le})}{V_t}+\frac{B_p}{M}\right)s+\frac{4\beta B_p(k_{qp}+C_{le})+4\beta A_A^2}{MV_t}}, \tag{13.28}$$

and recalling the standard second order form as:

$$G(s) = K\frac{\omega_n^2}{s^2+2\zeta\omega_n s+\omega_n^2}, \tag{13.29}$$

one may identify the system undampened natural frequency, damping and gain as:

$$\omega_n = \sqrt{4\beta\frac{k_{qp}B_p+C_{le}B_p+A_A^2}{MV_t}}, \tag{13.30}$$

$$\zeta = \frac{\left(\frac{4\beta(k_{qp}+C_{le})}{V_t}+\frac{B_p}{M}\right)}{2\sqrt{4\beta\frac{k_{qp}B_p+C_{le}B_p+A_A^2}{MV_t}}}, \tag{13.31}$$

$$K = k_{qx}\frac{4\beta A_A}{MV_t}\frac{MV_t}{4\beta B_p(k_{qp}+C_{le})+4\beta A_A^2} = k_{qx}\frac{A_A}{B_p(k_{qp}+C_{le})+A_A^2}. \tag{13.32}$$

If we set the supply pressure as input and assume the valve spool position and the load force to be constant, a system from supply pressure to velocity is formed:

$$\frac{sX_p(s)}{P_S(s)} = \frac{k_{qp}\frac{4\beta A_A}{MV_t}}{s^2+\left(\frac{4\beta(k_{qp}+C_{le})}{V_t}+\frac{B_p}{M}\right)s+\frac{4\beta B_p(k_{qp}+C_{le})+4\beta A_A^2}{MV_t}}. \tag{13.33}$$

Once again see that the velocity system is a second order system. Investigating influence of disturbance induced by changes in load force, one may derive a transfer function from load force to velocity which is:

$$\frac{sX_p(s)}{F_L(s)} = \frac{-(Ms+\frac{4\beta(k_{qp}+C_{le})}{MV_t})}{s^2+\left(\frac{4\beta(k_{qp}+C_{le})}{V_t}+\frac{B_p}{M}\right)s+\frac{4\beta B_p(k_{qp}+C_{le})+4\beta A_A^2}{MV_t}}. \tag{13.34}$$

Note that gain is negative from load force to cylinder piston velocity; this fact is evident from Fig. 11.4 where the load force is seen to work in the negative direction.

From the three transfer functions it is noted that increasing the spool position and supply pressure will increase the velocity. Conversely, increasing the load force will decrease the velocity. For both the spool position and the supply pressure a second order dynamic is clearly uncovered, where the load force includes a zero to the second order system.

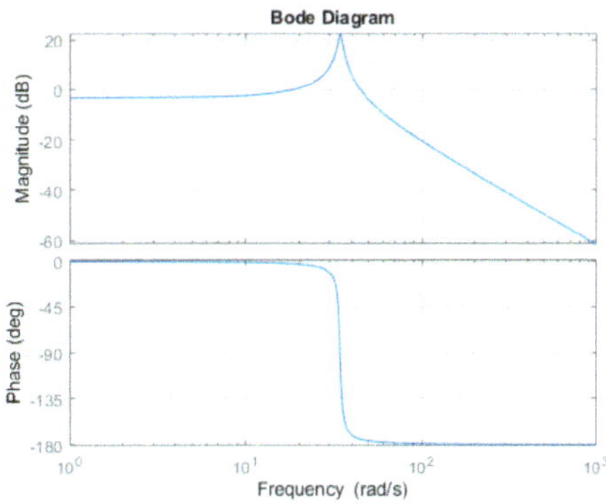

(a) Frequency response for the Cylinder-Valve Drive.

Symbol	Value	Unit
M	5000	kg
V_t	0.0015	m^3
B_p	4000	Bs/m
A_A	0.015	m^2
Q_{nom}	14	L/min
p_{nom}	10	bar
β	10000	bar
p_S	200	bar
p_{nom}	0	bar

(b) System parameter values.

Fig. 13.7 Frequency responses for the cylinder valve drive

13.3.2 Frequency Response

The frequency response for the reduced order model of the symmetrical cylinder-valve drive, (13.28) is seen in Fig. 13.7a for system parameters seen in Fig. 13.7b.

The frequency response clearly shows the second order system given by (13.28).

Parameter Variations

Let's use Bode diagrams to investigate how various parameters influence the frequency response. Four parameters are increased one by one and plotted from line 1–5 for increasing values. The investigated parameters are mass, bulk modulus, leakage coefficient and valve spool position in the linearisation point, with the latter strongly relating to the cylinder piston velocity at the linearisation point. The frequency responses in Fig. 13.8a, b reveals a significant effect on system eigenfrequency from change mass and bulk modulus, however little to no effect is seen on the damping and DC gain for both parameters.

When varying leakage and valve spool position the frequency response plot in Fig. 13.9a, b show a significant change in system damping but not in DC-gain and eigenfrequency.

13.3.3 System Understanding—Cylinder-Valve Drive

The frequency analysis clearly indicates that the cylinder-valve drive system is an "under dampened" second order system. It is, though, evident that one may expe-

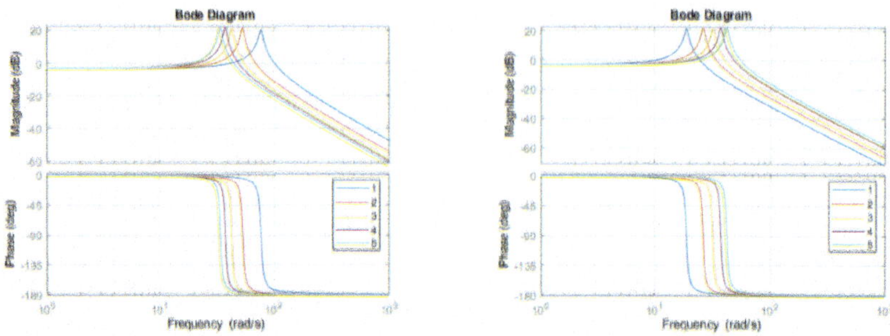

(a) Variation in mass. $M = [1000,\ 2000,\ 3000,\ 4000,\ 5000]$ kg.

(b) Variation in bulk modulus. $\beta = [3000,\ 6000,\ 9000,\ 12000,\ 15000]$ bar.

Fig. 13.8 Frequency responses to the cylinder valve drive

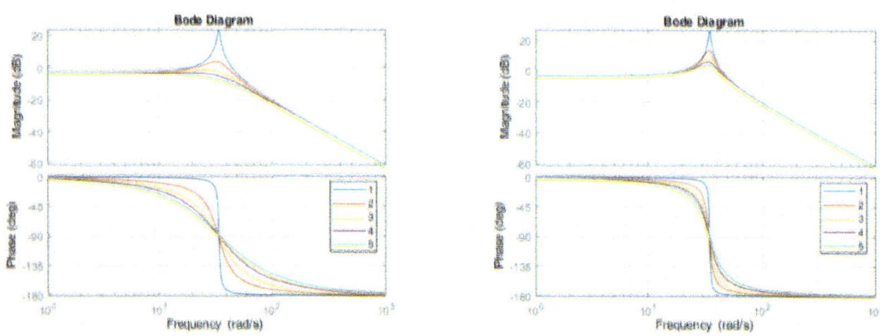

(a) Variation in leakage. $C_{le} = [0.0,\ 0.03,\ 0.06,\ 0.09,\ 0.12]\frac{\text{L}}{\text{min bar}}$

(b) Variation in valve spool position. $x_{v0} = [0,\ 5,\ 10,\ 15,\ 20]$ %

Fig. 13.9 Frequency responses in the cylinder valve drive

rience a dampened second order system if the leakage of the system is very high (unartistically high) or if one chooses a large valve opening area corresponding to a large cylinder piston velocity as linearisation point. It is evident that choice of linearisation point significantly influences controller design basis as one often develops controllers based on a frequency response; but which response should you base your design on?

Likewise, a significant change in system response is seen when the system load is changed. Imagine that the cylinder-valve drive is actuating e.g. a crane or excavator, two systems for which the load mass may very substantially. Should one design based on maximum, minimum or average mass?

Chapter 14
Control of Fluid Power Systems

Abstract This chapter introduces two system manipulation strategies highly applicable for fluid power systems. Firstly, active damping by pressure feedback, both direct and high pass filtered pressure feedback is shown to significantly increase system damping. Secondly, a valve compensation strategy eliminating the valve flow non-linearity is given. In addition to the two "manipulation" strategies, flow feedforward is introduced as an effective add-on when performing reference tracking. Finally, all these add-on strategies are combined with standard linear controller strategies.

Even though fluid power systems are non-linear in nature (especially valve-cylinder drives), linear controllers have shown themselves to be feasible in most control cases of such systems. Figure 14.1 illustrates a standard approach to linear control, with a system plant, $G_\mathrm{p}(s)$, a feedback loop, $H(s)$, and a controller, $G_\mathrm{c}(s)$, acting on the error, $e(s)$.

In valve controlled fluid power systems the control input is most often valve spool position, hence why the controller output is a valve reference. In Fig. 14.2 a simple position feedback control of a symmetric cylinder-valve drive with constant supply pressure, such as the one modelled and analysed in Sects. 11.3 and 13.3, is depicted.

Likewise, velocity feedback control of the same symmetric cylinder-valve drive with constant supply pressure is seen in Fig. 14.3.

In both the position and velocity system a controller $G_\mathrm{c}(s)$ is to be designed. Such a basic controller design is untouched in the current version of this textbook. We shall only note that the output of the controllers is the reference for the valve such that:

$$G_\mathrm{c}(s) = \frac{X_\mathrm{v}(s)}{E(s)}. \tag{14.1}$$

Note that the error $E(s)$ depends on the type of feedback and input reference (i.e. position, velocity, pressure).

In the remaining parts of this chapter, I will firstly study how one may add to the valve reference generated in the basic controller, $G_\mathrm{c}(s)$, such the new control law will be:

A. H. Hansen, *Fluid Power Systems*, Fluid Mechanics and Its Applications 129,
https://doi.org/10.1007/978-3-031-15089-0_14

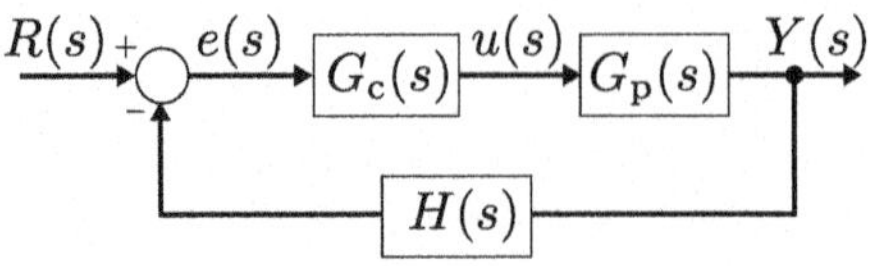

Fig. 14.1 Standard linear control system

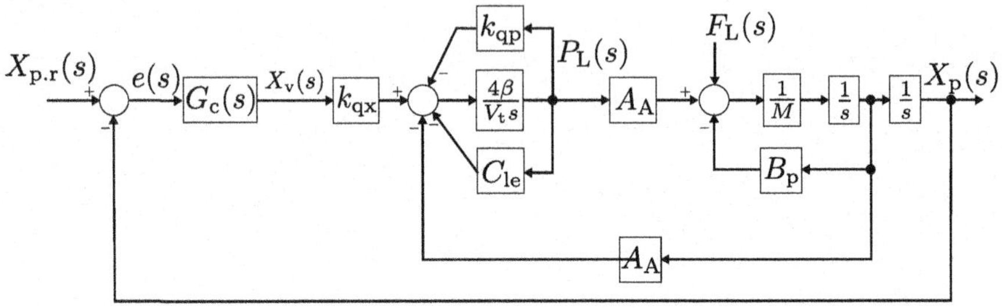

Fig. 14.2 Block diagram for a symmetric cylinder with feedback position control

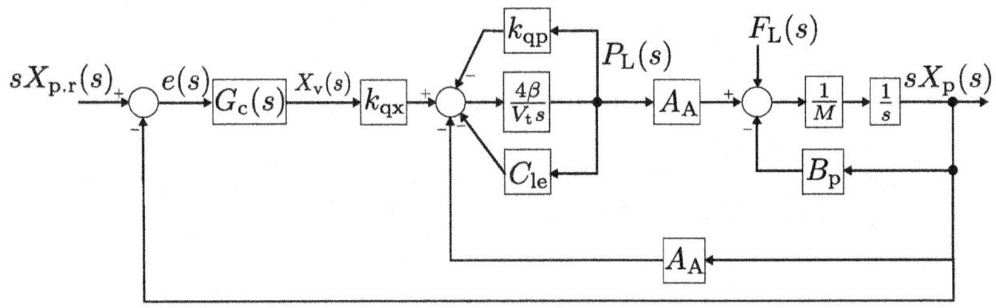

Fig. 14.3 Block diagram for symmetric cylinder with velocity controller

$$x_{\mathrm{v.r}}(t) = G_{\mathrm{c}}e(t) + x_{\mathrm{v.rp}}(t) + x_{\mathrm{v.rf}}(t), \tag{14.2}$$

where $x_{\mathrm{v.r}}$ is the valve reference given to the system from the "full" controller. The first term $G_{\mathrm{c}}e(t)$ is the valve reference generated by the basic feedback controller e.g. velocity controller. The second $x_{\mathrm{v.rp}}(t)$ and third $x_{\mathrm{v.rf}}(t)$ term are additions to the valve reference given by a pressure feedback loop and a velocity feed forward respectively. We shall discuss how system damping may be increased with a pressure feedback loop and how velocity feed forward may ease the task of the tracking controller, $G_{\mathrm{c}}(s)$.

Secondly, a section is dedicated to studying how a pre-compensator may be implemented to eliminate the non-linearity inherently induced by the orifice equation in valve controlled systems. This will significantly simplify the controller design but may also pose some stability issues to be aware of.

The chapter will close with a brief discussion of multiple input multiple output systems, and an example of a Separate Metering in Separate Metering Out system will be illustrated.

14.1 Pressure Feedback

As pressure measurements are relatively cheap and often at hand in fluid power systems, pressure feedback may be used in the control algorithms. When focusing on linear control algorithms a direct pressure feedback loop may yield an add-on to the control law as:

$$x_{\mathrm{v.rp}} = k p_{\mathrm{L}}. \tag{14.3}$$

Let's define the controller constant such that it counts in the valve flow linearisation constant, k_{qx}, such that:

$$x_{\mathrm{v.rp}} = -\frac{k_{\mathrm{p}.p_{\mathrm{L}}}}{k_{\mathrm{qx}}} p_{\mathrm{L}}. \tag{14.4}$$

Note that if implemented with an ideal valve model, this yields a feedback loop parallel with the internal leakage in the cylinder. Recalling the influence of the leakage on the system dynamics—see Bode diagram in Fig. 13.9—one may note that a negative pressure feedback loop will increase system damping—however, at the cost of a lower system gain. The philosophy of the pressure feedback loop is to lower the valve reference signal as the load pressure increases and visa versa. This seeks to eliminate changes in the load pressure, however not all load pressure changes are unwanted. As high frequency changes are often unwanted, low frequency changes may be required to perform the intended movement of the actuator.

To allow the low frequency changes while still suppressing high frequency pressure changes/oscillations, a high pass filter may be applied in the pressure feedback loop. The control law of the pressure feedback loop is then given as:

$$x_{\mathrm{v.rp}} = -\frac{k_{\mathrm{hp}.p_{\mathrm{L}}}}{k_{\mathrm{qx}}} \frac{s}{\tau_{\mathrm{hp}} s + 1} \cdot p_{\mathrm{L}} \tag{14.5}$$

Performing a system analysis with the pressure feedback laws in (14.4) and (14.5) implemented with an ideal proportional-valve will yield the Bode diagrams in Fig. 14.4 showing the open loop velocity system when the controller is a proportional controller with gain 1 ($G_{\mathrm{c}}(s) = 1$). G_p is for the system with only a proportional controller given as ($G_{\mathrm{c}}(s) = 1$) and no pressure feedback.

Note the difference in the DC-gain for the pressure (G_p.p) and high pass pressure feedback(G_p.ph), where the pressure feedback yields a lower system gain compared to the standard system. The high pass pressure feedback compensated system has equal DC-gain as the standard uncompensated system. Both the high-pass pressure and the pressure feedback yield an increased system damping.

With one of these pressure compensation strategies implemented the design engineer is given a much simpler system for which to design a standard control, $G_t exts(s)$.

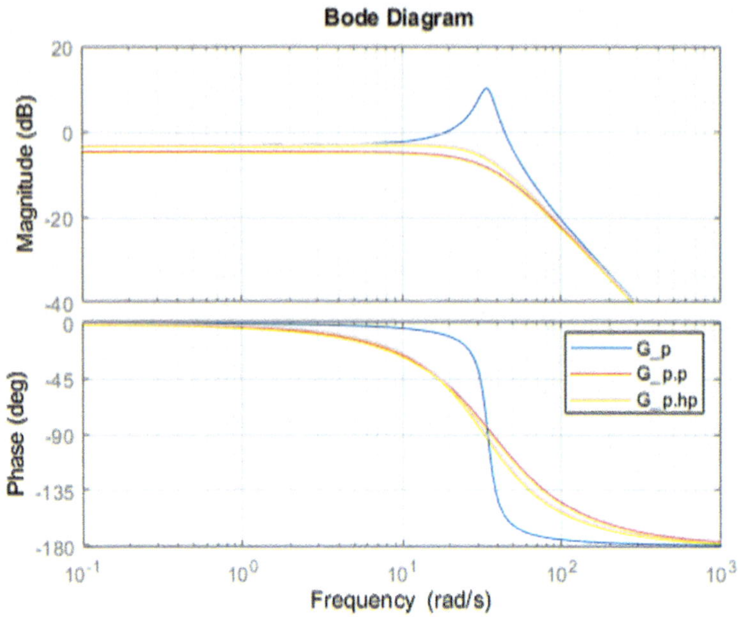

Fig. 14.4 Bode diagram of system plant, standard, with pressure and high pass pressure feedback. $k_{\text{p}.p_{\text{L}}} = 20C_{\text{le}}$, $k_{\text{hp}.p_{\text{L}}} = 3C_{\text{le}}$, $\tau_{\text{hp}} = \frac{1}{5}$

14.2 Flow Feed Forward

When a system is controlled to follow a trajectory that is known some time in advance, flow feed forward may be applied. Flow feed forward is an algorithm where the velocity trajectory is used as input to an open loop velocity control, i.e. the valve spool is given a reference such that it matches the required flow; however, no velocity or flow feedback is used.

If pressure feedback is utilised to adjust the gain, the flow feed forward algorithm is named active flow feed forward. By contrast, the algorithm using only the velocity reference and no feed back is named passive flow feed forward.

The control law for flow feed forward is given based on the desire to compensate the displacement flow due to piston movement ($Q = \dot{x}_{\text{p}} A_{\text{A}}$). From this, the valve reference from the velocity feed forward is, in the time domain, given as:

$$x_{\text{v.rp}}(t) = \frac{\dot{x}_{\text{p}}(t) A_{\text{A}}}{C_{\text{d}} w \sqrt{\frac{1}{\rho}(p_{\text{S}}(t) - p_{\text{L}}(t))}}. \tag{14.6}$$

As flow feed forward is an open loop control it will most often be implemented together with a feedback tracking loop. This feedback tracking controller, however, now only needs to compensate for the error after the open loop controller as worked and as such the control effort from the feedback controller $G_c(s)$ is much smaller and the controller gain may therefore often be increased to gain better tracking performance and disturbance rejection.

14.2.1 Passive

In the passive algorithm the pressure difference in the control law is a best estimate of the pressure difference, such that the control law becomes:

$$x_{v.rp}(t) = \frac{\dot{x}_{p.r}(t) A_A}{C_d w \sqrt{\frac{1}{\rho} \Delta p_v}}, \tag{14.7}$$

where Δp_v is a constant value and $\dot{x}_{p.r}$ is the velocity reference. Naturally the pressure drop across the valve is not constant. The more the pressure differs from the estimate, the worse the flow feed forward algorithm compensates the displacement flow. One may note that no state feedback is introduced and as such no system dynamics (poles) are changed.

14.2.2 Active

Pressure measurements are available in many fluid power systems, and may as such be used in flow feed forward control laws. In case of the reduced order model for the symmetric cylinder-valve drive (see Sect. 11.3) the flow feed forward may be given as:

$$x_{v.rp}(t) = \frac{\dot{x}_{p.r}(t) A_A}{C_d w \sqrt{\frac{1}{\rho} (p_S(t) - p_L(t))}}. \tag{14.8}$$

To analyse the effect of flow feed forward on the system dynamics the control law is linearised to (14.9):

$$X_{v.rp}(s) = \frac{1}{k_{qx}} \left(s X_{p.r}(s) A_A - k_{qp} P_S(s) + k_{qp} P_L(s) \right). \tag{14.9}$$

In Fig. 14.5 the flow feed forward is included with dotted lines. Note that the flow feed forward valve reference is added to the reference of a cylinder piston position controller.

As the active flow feed forward uses the pressure measurements, the compensation is more precise. If the compensation is ideal the block diagram will be reduced to the one seen in Fig. 14.6. Note that the displacement flow is cancelled out as well as the valve flow dependency on load and supply pressure (in the orifice equation).

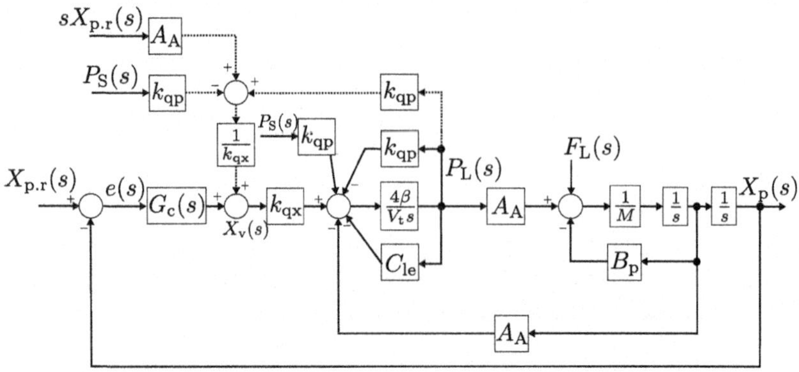

Fig. 14.5 Block diagram of position controlled symmetric cylinder with active flow feed forward

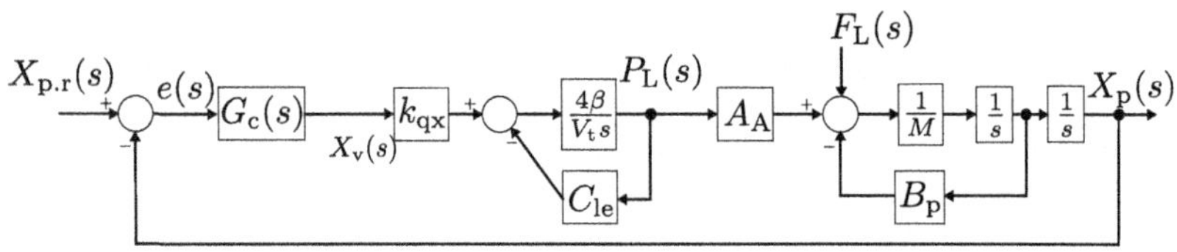

Fig. 14.6 Reduced block diagram when an ideal flow feed forward is used

When the compensation is perfect this implementation gives good results. However, as the valve dynamic is unmodelled and some modelling imprecision may exist the cancellation will never be perfect.

The compensation of the load pressure will lower the system damping. This is seen from the induced positive feedback loop that cancels the negative pressure feedback through k_{qp} which added damping. If the load pressure increases, the valve flow would normally decrease due to lower pressure drop; however, the flow feed forward compensates for this by increasing the valve opening. If the valve model is imprecise such that the k_{qp} value in the flow feed forward is larger than the system value, a positive feedback loop is made, possibly leading to instability or oscillations as increasing load pressure leads to larger valve opening, yielding more flow which may increase the load pressure further.

To avoid the negative effect of compensating for the pressure difference, it is beneficial to consider which pressure changes to compensate for. It is often the low frequency pressure changes that the flow feed forward should compensate for, as these are within the system dynamic range. Conversely, the natural damping effect of the valve is wanted for oscillations in the region of the system eigenfrequency and higher.

14.3 Valve Compensator

A main contributor to non-linearity in hydraulic systems is orifice flows. This is especially relevant in valve controlled systems such as those treated in this note. For many systems the input is a valve spool reference to a proportional valve, the valve spool position together with an actual pressure drop yield a given flow. A valve compensator is designed such that a flow reference may be given as input rather than a valve spool position. The valve compensator is in fact the gain for a flow feed forward. Such a valve compensator is given as:

$$x_{v.ref}(t) = \frac{Q_{ref}(t)}{k_{vE}\sqrt{\Delta p_{meas}(t)}}. \tag{14.10}$$

where k_{vE} is an estimate of the valve flow gain. Note, that the valve compensator is the inverse of the orifice equation. One must realise that the valve coefficient and the pressure drop must be the one for the valve path controlling the flow associated with Q_{ref}, (this is relevant for e.g. 4/3 way proportional valves).

Inserting the valve compensator in the orifice equation yields:

$$Q(t) = k_v x_v(t)\sqrt{\Delta p}, \tag{14.11}$$

$$Q(t) = Q_{ref}(t)\frac{k_v}{k_{vE}}\frac{\sqrt{\Delta p}}{\sqrt{\Delta p_{meas}(t)}}. \tag{14.12}$$

when assuming $x_v(t) = x_{v.ref}(t)$, hence the valve is assumed infinitely fast. It is clearly seen that a perfect valve coefficient estimation k_{vE} and pressure measurement leads to the actual flow being equal to the flow reference, at least this is the case in steady state.

$$Q(t) = Q_{ref}(t). \tag{14.13}$$

With the valve coefficient and the pressure measurement on target the relation of valve spool reference and valve spool position is of interest. A linear model of the valve dynamic may be included to account for non-ideal valve dynamics, such that the dynamic relation between the flow reference and the actual flow is given by the valve dynamics.

$$Q(t) = \frac{x_v(s)}{x_{v.ref}(s)} Q_{ref}(t). \tag{14.14}$$

With this valve compensator the engineer may disregard the valve model in the system modelling and use flow reference as input to the model. One may note that the non-linear nature of the orifice equation is compensated.

14.4 Valve Dynamics

Throughout most of the system modelling and analysis discussion in the current note the valve dynamics have been neglected for simplicity. However, little investigation of the validity of such assumption has been conducted.

When designing controllers and system manipulators, the design engineer needs to consider the dynamics of the valve, electrical motor or whichever actuator is given a control input. Especially for the high pass pressure feedback loop, the valve dynamics are important, as pressure oscillation in the range of system eigenfrequency has to be compensated for. For this to work properly the valve is required to be significantly faster than system eigenfrequency.

The valve dynamics may naturally be included in the modelling, however, at the cost of a higher order system complicating the easy system overview. Recall that the reduced order system yielded a second order system with the free integrator for cylinder piston position.

14.5 Multi-input Systems

When dealing with fluid power systems, the full system often has more than one input even thought the main objective is e.g. velocity control. Hence, in general the system has multiple inputs and maybe only one primary output, e.g. cylinder piston velocity.

Let's think of the symmetrical fluid power cylinder controlled with a servo valve. The linear system model (11.64) has four inputs, $u = [p_S \; p_T \; x_v \; F]^T$. Let's assume tank pressure and load force to be constant and assume the supply pressure to be controllable, e.g. $\frac{p_S}{p_{S,ref}}$. This could be a typical fluid power system having two inputs, (e.g. pump swash plate angle and valve spool position) and one primary output requested to follow some reference.

A detailed treatment of multi-variable control is beyond the scope of this note. However, the fluid power engineer has to consider which states should be controlled. For the system with supply pressure and valve spool position as input one could set the valve fully open for positive piston velocity and then control the speed with the pump pressure. Or the pump pressure could be set to some value larger than the load pressure such that it is controlled according to the load force requested (load sensing). Naturally the supply pressure could be set to a constant high value e.g. 250 bar, yielding only one variable input; however, at the cost of a large energy usage.

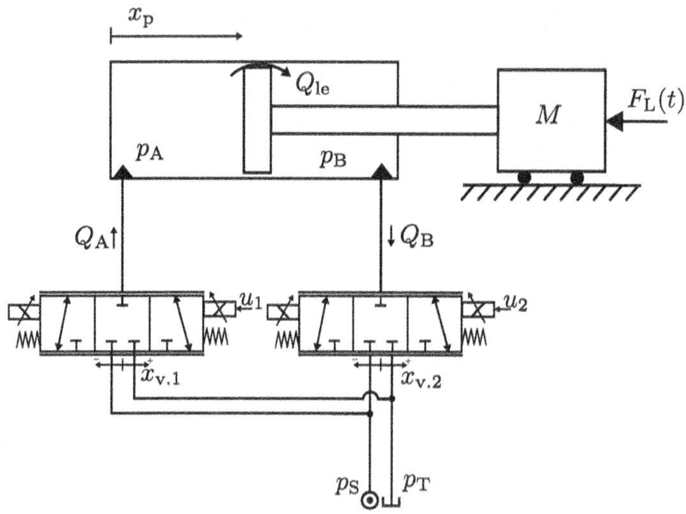

Fig. 14.7 Illustration of a SMISMO system

14.5.1 SMISMO—System

Separate meter in separate meter out systems are another system type featuring at least two inputs. Let's imagine a system consisting of a differential cylinder controlled with two 3/3 way proportional valves, each connected to a fixed supply and tank pressure. In such a system the flow edge to each cylinder chamber is controlled separately and has the inputs, u_1 and u_2 (Fig. 14.7).

The controller designer may use one valve to control cylinder piston velocity and another valve to control pressure level in the cylinder chambers. However, the coupling between inputs and outputs may require a further study and some measures to be made.

Chapter 15
Fluid Power Systems Design

Abstract This chapter introduces a cookbook method for system design. The method is based on steady-state analysis, and each step is given with simple formulas for component sizing.

Throughout the lecture note, the focus has been on modelling, analysing and controlling existing systems. In this chapter, the focus is directed at how to design a fluid power system such that it will fulfill a given task. How is a system layout chosen? Which components should one use? How should the system be controlled? These are some of the important questions to answer through the design process. In this chapter, a 14 step "cookbook" algorithm for system design is presented. The steps are:

1. Define operation of the system.
2. Define operation cycles for each sub-function in the system.
3. Define system architecture and construct a fluid power diagram.
4. Select operating pressure for the system.
5. Compute and select actuator size.
6. Compute and select pump and primary mover size.
7. Select type of fluid.
8. Select size of the fluid lines.
9. Select control elements and settings.
10. Perform a steady state analysis of the system, and determine overall efficiency.
11. Select and size tank and cooling systems.
12. Define filtration requirements.
13. Prescribe procedures for system assembly, monitoring, operation and component replacement.
14. Estimate costs.

Each step is introduced in the following 14 subsections, such that after completing this chapter the reader will be left with tools to design their own fluid power system. The algorithm presented is based on a steady state approach to system design, therefore it may, depending on system tasks, be necessary to evaluate how the system performs dynamically.

A. H. Hansen, *Fluid Power Systems*, Fluid Mechanics and Its Applications 129,
https://doi.org/10.1007/978-3-031-15089-0_15

15.1 System Operation

For any model based design procedure the function of the system is the design based point. Therefore, the starting point of a design procedure is to define the overall function of the system and operating cycle of the system. This function description serves as an overview of the design and system specification. Furthermore, subfunctions may be derived from the overall system requirements.

15.2 Operation of Subfunction

Based on the overall system operating cycle, subfunction operating cycles are defined. This will be on actuator level, e.g. a required velocity and force for a linear motion.

With well defined subfunction operating cycles the system designer has well defined system requirements. Furthermore, the operating cycles may, together with proper documentation during the design phase, improve traceability in the developments of systems.

The operating cycle may be described in various levels of detail depending on how precisely the mechanical structure of the system is known. The more well defined the mechanical structure and loading situation the more precisely the operating cycle can be defined for the subfunctions.

However, it is recommended to set up time series operating cycle diagrams for all actuators, in terms of load (force/torque) and velocity (linear/angular). An illustration of such is seen in Fig. 15.1.

15.3 System Architecture—Diagram

This task of "System Architecture" is twofold; initially, the system architecture is defined. This is basically to determine how fluid power shall be generated and used to perform the intended work, i.e. how shall chemical or electrical energy be converted to fluid power energy and how shall this fluid power energy perform a mechanical work? Four points to consider in defining the system architecture are:

- How many pumps will be used?
- How are the pumps driven and controlled?
- How many and which types of actuators will be used?
- How should each actuator be controlled?

Based on answers to the above questions an initial fluid power diagram may be sketched.

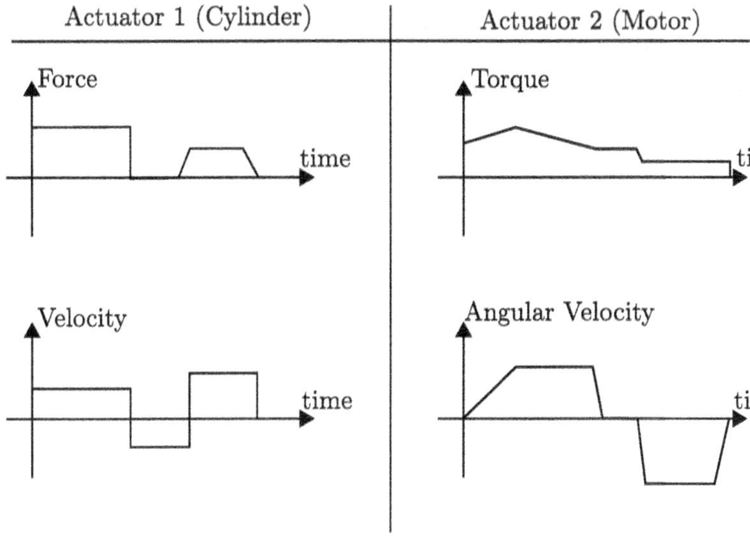

Fig. 15.1 Illustration of subfunction operating cycle diagram

15.4 System Pressure Level

Before the sizing of components, an operational pressure level is chosen. The pressure level often lies within 150–350 bar. However, if a pressure level in this interval yields very low or high flows (less than 3–5$\frac{\text{L}}{\text{min}}$ or above 1000$\frac{\text{L}}{\text{min}}$) the pressure level may need to be lower or higher respectively. This naturally depends on the given system tasks; in high power systems 1000 $\frac{\text{L}}{\text{min}}$ may be a feasible flow level.

15.4.1 Actuator Sizing

With an architecture and pressure level to hand one may start sizing actuators. The task is to calculate required motor displacement and cylinder piston areas.

Motor displacement may be determined from:

$$D_{\text{m}} > \frac{2\pi T_{\text{max}}}{p}, \tag{15.1}$$

where,
D_{m} Displacement $[\frac{\text{m}^3}{\text{rev}}]$
T_{max} Maximum torque on the motor [Nm]
p Pressure level [Pa].

Note that a lower limit for the displacement is found and so the displacement of the chosen motor must be larger than the calculated value.

Cylinder piston area may be determined by:

$$A > \frac{F_{\max}}{p}, \tag{15.2}$$

where;
A Piston area [m^2]
$F_{\max}$ Maximum force on the piston [N]
p Pressure level [Pa].

As for the motor, the calculated value is a lower bound for the piston area.

In these calculations, no friction, i.e. hydro-mechanical efficiency, has been included. It may be necessary to choose a somewhat larger motor displacement and cylinder piston area respectively. One must likewise consider the pressure level use; is it the operational pressure level at pump outlet? If so, pressure losses through pipes, valves etc. should be considered and the pressure level used could be chosen 20–40 bar below the pump operational pressure level.

15.5 Pump and Primary Mover Sizing

With actuator sizes and velocity time series one may calculate the flow demand. The flow is supplied from the pump(s) in the system and one may now size the pump system. The total maximum flow request is given as the sum of maximum flow demanded from each actuator:

$$Q_{\max} = \left(\sum_{i_m}^{N_m} D_{m.i} n_{m.i} + \sum_{i_c}^{N_c} A_{A.i} \dot{x}_{p.i} \right)_{\max}, \tag{15.3}$$

where;
$Q_{\max}$ Maximum total consumer flow [$\frac{m^3}{s}$]
$D_{m.i}$ Displacement of i'th motor [$\frac{m^3}{rev}$]
$n_{m.i}$ Rotational speed of i'th motor [$\frac{rev}{s}$]
$A_{A.i}$ Piston area of i'th cylinder [m^2]
$\dot{x}_{p.i}$ Speed of i'th piston [$\frac{m}{s}$].

A well defined operation cycle will enable price calculation of maximum flow demands, as timing between operations may be taken into account. One may yield rather different maximum flow demands if one solely measures maximum flow demands from each actuator or if one summarises the flow times series based on an operation cycle and takes maximum flow request from the combined flow time series.

If the rotational velocity of the primary mover is known, the pump displacement is given as:

$$D_{\mathrm{p}} > \frac{Q_{\max}}{n_{\mathrm{p}}}, \tag{15.4}$$

where;
D_{p} Pump displacement $[\frac{\mathrm{m}^3}{\mathrm{rev}}]$
$Q_{\max}$ Maximum total consumer flow $[\frac{\mathrm{m}^3}{\mathrm{s}}]$
n_{p} Rotational speed of pump $[\frac{rev}{s}]$.

With pump displacement chosen and pressure level known, the torque demand on the primary mover is given as:

$$T_{\mathrm{p}} > \frac{D_{\mathrm{p}}p}{2\pi}, \tag{15.5}$$

where;
T_{p} Torque required for the pump [Nm]
D_{p} Pump displacement $[\frac{\mathrm{m}^3}{\mathrm{rev}}]$
p Pressure level [Pa].

In the above calculations no efficiencies are included. The "greater than" sign should be used to account for e.g. actuator and pump volumetric efficiency leading to a higher maximum flow. Furthermore, the hydro-mechanical efficiency of the pump will bring a higher torque requirement on the primary mover.

The maximum power requirement on the primary mover driving the pump may be found from the maximum flow and the pressure level, as:

$$\mathcal{P}_{\max} > Q_{\max}\, p, \tag{15.6}$$

where;
$\mathcal{P}_{\max}$ Maximum total consumer power [W]
$Q_{\max}$ Maximum total consumer flow $[\frac{\mathrm{m}^3}{\mathrm{s}}]$
p Pressure level [Pa].

15.6 Fluid

When designing a fluid power system, the choice of working fluid is not negligible as both system functions and lifetime may be greatly influenced by the fluid used. Numerous aspects have to be accounted for when choosing the working fluid, i.e.:

Table 15.1 Characteristics for a Shell Tellus mineral oil [2]

Shell Tellus oil	22	32	37	46	68	100
ISO oil type	HM	HM	HM	HM	HM	HM
Kinematic viscosity, ν						
0 °C cSt	180	338	440	580	1040	1790
40 °C cSt	22	32	37	46	68	100
100 °C cSt	4.3	5.4	5.9	6.7	8.6	11.1
Viscosity index	100	99	99	98	97	96
Density @15 °C	866	875	875	879	886	891
Flash Point°C	204	209	212	218	223	234
Pour Point°C	–30	–30	–30	–30	–24	–24

- Required viscosity.
- Lubrication ability.
- Working temperature.
- Fire risk.
- Environmental conditions.

Mineral oil is still the most used fluid in fluid power systems. However, oil-in-water or water-in-oil emulsion, phosphate ester based fluids and biodegradable hydraulic fluids are getting more focus. For all fluid types, additives are used to get certain properties, e.g. additives for oxidation protection, anti-foam, improved lubrication etc.

When choosing fluid for a fluid power system the design engineer must thus consider both the normal operation condition and break down. It is advisable to contact the producer of the fluid to get the right information.

Table 15.1 shows some common oils supplied by Shell. In the table, various fluid parameters are listed with viscosity and density being the most important with regard to system modelling. Note that this is supplier data and not the models discussed in Sect. 2.1.

15.7 Fluid Lines

Fluid lines in a fluid power system consist of pipes and hoses as well as various fittings. When choosing pipes and hoses the engineer needs firstly to ensure that the burst and allowable operation pressure of the pipes and hoses are beyond the pressure level chosen. Furthermore, the dimensions of the pipes and hoses should be chosen such that a feasible fluid velocity is experienced in the lines. With high flow velocities comes high pressure loss, however with low velocity comes larger pipe diameters which is costly and requires larger material thickness. As a rule of thumb the fluid velocity should lie in the region of:

- Suction line: 0.5 – 2.0 $[\frac{m}{s}]$ (Tank to pump)
- Delivery line: 3.0 – 8.0 $[\frac{m}{s}]$ (Pump to actuator)
- Return line: 1.0 – 3.0 $[\frac{m}{s}]$ (Actuator to tank)

Even when obeying these rules it might be necessary to check whether pressure drops in lines of the system are within reasonable values. In particular, the pressure drop in the suction line may be critical as the pressure head is low and cavitation at the pump inlet may damage the pump. Note that the tank pressure is often only 1 bar absolute pressure.

15.8 Control Elements

System control elements are chosen according to the system diagram constructed in Sect. 15.3 in terms of functionality. Often it is preferable to place control valves together and close to either the pump or the actuator that each valve is controlling. Valves may be plate mounted, block mounted or some other combinations. However, regardless of the valve type it is often beneficial to use valve mounts with the same connection size as the chosen line size. Hence, if a 1/2 "pipe is chosen, a valve with a 1/2" connection is preferred in order that no additional pressure losses are introduced in the transition from line to valve.

15.9 Steady State Analysis and Overall Efficiency

At this point in the design procedure all components important to system function have been chosen, operation cycles are defined and control element settings decided. With the preliminary design at hand a system analysis is performed according to Chap. 12. With the result of a steady state analysis the design engineer can evaluate system functionality.

The overall system efficiency may in addition be calculated as:

$$\eta_{\text{tot}} = \int_{t=0}^{t=t_c} \frac{\mathcal{P}_{\text{out}}}{\mathcal{P}_{\text{in}}} dt, \tag{15.7}$$

where;
η_{tot} Total system efficiency [–]
$\mathcal{P}_{\text{out}}$ power delivered by the system actuators [W]
$\mathcal{P}_{\text{in}}$ power put into system [W]
t_c duration of the operation cycle for the system [s].

The efficiency indicates that losses appear in the system. The energy lost in the system is the difference between input energy and output work. The losses are due to non-ideal components, but as seen in Chap. 12 energy is additionally lost in the control elements depending on system architecture and sizing of components.

15.10 Tank and Cooling

Energy Losses in the system will lead to heating of the fluid. In general, losses in the system lead to heat generation and mostly heating of the fluid, however, as fluid passes through system components and pipes, these are heated as well, and some power is dissipated in the heating of components and from these components to the surroundings. Let's as a conservative approach assume that energy is only dissipated from the fluid in the tank, and let's furthermore assume the fluid temperature to be uniform across the entire fluid in the system.

Heat flux generated in the system and heat flux from the tank to its surroundings are then given as:

$$\Phi_s = \mathcal{P}_{in} - \mathcal{P}_{out} = (1 - \eta_{tot}) \cdot 2\pi \cdot T_e \cdot n_p, \tag{15.8}$$

$$\Phi_r = k_c A_r \Delta T, \tag{15.9}$$

where;
Φ_s Heat flux generated by the system [W]
Φ_r Heat flux transmitted to the surroundings [W]
$\mathcal{P}_{in}$ power put into the system [W]
$\mathcal{P}_{out}$ power delivered by the system actuators [W]
η_{tot} Total system efficiency [–]
k_c Heat transmission coefficient for the tank $\left[\frac{W}{m^2K}\right]$
A_r Cooling area of the tank [m^2]
ΔT Temperature difference between fluid and surroundings [K].

In the case of an imbalance between heat flux generated and transmitted to the surroundings, the fluid temperature will change. However, it is desirable to keep the fluid temperature constant as too high a temperature will damage the fluid and too low a temperature will yield high fluid viscosity. Cooling is often required in fluid power systems, and also heating of the fluid may be necessary in some cases.

15.11 Filtration

Failures in properly designed fluid power systems are mostly due to contamination of the fluid. As a consequence, filtering of the working fluid is of high importance if a proper working system is desired. Maximum contamination level is a data sheet information for most fluid power components, such as pumps, motors and valves. Three filter placements are commonly used, e.g.

- Suction filter.
- Pressure filter.
- Return filter.

Table 15.2 Table for ISO 4406, oil contamination

Code	More than	Up to/including
24	80000	160000
23	40000	80000
22	20000	40000
21	10000	20000
20	5000	10000
19	2500	5000
18	1300	2500
17	640	1300
16	320	640
15	160	320
14	80	160
13	40	80
12	20	40
11	10	20
10	5	10
9	2.5	5
8	1.3	2.5
7	0.64	1.3
6	0.32	0.64

The suction filter is placed between tank and pump, i.e. this filter protects the pump and components after the pump. The suction filter must be made with a low pressure drop so that cavitation is avoided. A pressure filter is typically placed after the pump, hence protecting components after the pump, especially valves. As it is placed in the high pressure line, the filter is dimensioned to high pressure levels. Compared to the suction filter, the pressure filter also protects valves from contamination from within the pump. A return filter is placed in the return line just before the tank, which is why this filter removes any contamination picked up by the fluid throughout the system. Return filters experience only low pressure levels and may as such be designed "weaker" and cheaper than a pressure filter.

The required filter efficiency is defined by system components. The filtration efficiency is given as two or three numbers and an ISO code, e.g. in [1] the Moog D663 reads ISO 4406 $>$ 17/14/11 and in [3] the Parker 4D01 reads ISO 4406 18/16/13. The ISO 4406 code numbers are given in Table 15.2. Let's derive the particle level allowed for the Parker directional valve in Table 15.3 .

The current valve allows for up to: 2500 particles per mL with a size of up to 4 μm; 1300 particles per mL with a size of up to 6 μm; and 160 particles per mL with a size of up to 14 μm. This is a rather relaxed demand on the filtration which

Table 15.3 Allowed particle level for the Parker 4D01

Particle size (μm)	ISO code	Allowed particles per mL
4	18	2500
6	17	1300
14	13	160

Table 15.4 Filter efficiency

Particle after filter	β-value at x μm	Filter efficiency (%)
50000 100000/50000	2	50
5000 100000/5000	20	95
1333 100000/1333	75	98.70
1000 100000/1000	100	99
500 100000/500	200	99.50
100 100000/100	1000	99.9

is typical for directional valves; in contrast, servo valves normally impose higher filtration demands.

As mentioned above, filters are designed for various pressure levels and flow rates. In addition, filters are given an efficiency code which is written as $\beta_x \geq \frac{N}{M}$. The β value is given for particle size x, and N is the number of particles larger than x flowing to the filter and M is the number of particles larger than x leaving the filter. Assuming that 100'000 particles are flowing towards the filter and after the filter, this number is reduced to 500. This results in a β value of 200.

When choosing a filter then, the designer first chooses the particle size in terms of the x value and then the efficiencies. An overview of the filter efficiency is given in Table 15.4.

Hence a $\beta_{10} \geq 75$ filter filters away 98.7% of particles greater then 10 μm.

15.12 Procedure for Assembly, Operation and Maintenance

With a final design inform of fluid power diagram, component list and controller settings it is time to consider practical issues such as assembly, operation and maintenance. How to assemble the system? How should the system be operated and monitored? What is the maintenance interval and what are the tasks?

In model based design we seldom consider these practical issues. It is, however, vital that our designed system can be assembled. It may be reasonable to construct a 3D model of the system containing all components including pipes, hoses and fittings, in order to evaluate whether enough space is available for tools and lifting equipment.

Together with a system design, an operation description documents how the system is to be operated to obtain the requested performance. This should include which type and how often maintenance is required, e.g. how often should oil and filters be changed. What should be monitored to detect possible fails in the system?

15.13 Estimate Costs

When designing machines costs are most often a relevant factor that need consideration in the design phase. If one can design a cheaper system with similar functionality this is often desired from customers. To estimate cost three main contributors must be included. Firstly, the raw component cost, which is the summarised cost of the components constituting the system. Secondly, assembly cost. Depending on system design and size this item may vary quite substantially. Lastly, operation and maintenance cost. This is the cost of the requested energy; an energy efficient system may be more costly to build but much cheaper to run.

References

1. Servovalves direct drive servovalves D633/634 series, Rev 2. 04/2009. Moog Inc
2. https://www.shell.com/
3. Catalog HY11-3500/UK, directional control valve, series 4D01. Parker Hannifin Corporation

Part IV
Learning by Doing

This last part of the Lecture Note contains exercises and solutions. The author truly believes that to master the topics within this note, it is vital to work with the exercises and reflect on the methods and results. The exercise chapter is ordered to follow the chapters of the notes, so for each chapter an exercises section can be found.

Chapter 16
Exercises

Abstract This chapter holds training exercises such that the student of fluid power systems may apply theory and formulas from the lecture note. The exercises are closely coupled to the theory presented in the note.

16.1 Fluid Mechanics I

16.1.1 Fluid Compressibility

Given a tank with a fixed volume of 100 L and a fluid bulk modulus, β, of 10000 bar, how much fluid must be added to the tank to raise the pressure 100 bar? See Fig. 16.1.

16.1.2 Fluid Spring

A fluid spring is seen in Fig. 16.2. It consists of a cylinder barrel and a piston rod. When the rod is at position $x = 0$ the fluid volume is $V = D^2 \frac{\pi}{4} L$. A constant fluid bulk modulus is assumed $\beta_f = 15000$ bar. $D = 100$ mm, $d = 30$ mm and $L = 500$ mm.

At point 1 $x_1 = 0$ the pressure is 10 bar.

(a) Derive an expression for the fluid volume as a function of piston position, $V(x)$.
(b) Calculate the force at point one, $F(x_1)$.

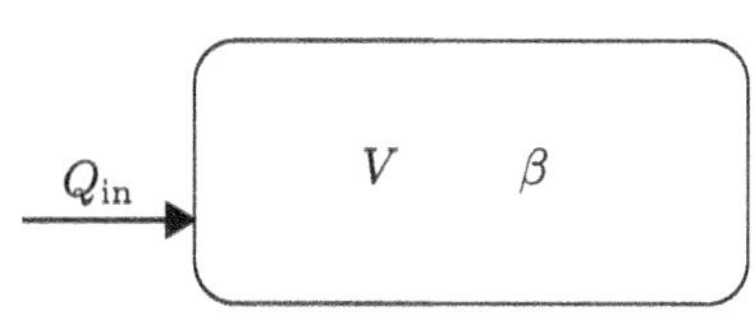

Fig. 16.1 Illustration of a fixed volume with an in flow

A. H. Hansen, *Fluid Power Systems*, Fluid Mechanics and Its Applications 129,
https://doi.org/10.1007/978-3-031-15089-0_16

Fig. 16.2 Illustration to exercises Sect. 16.1.2

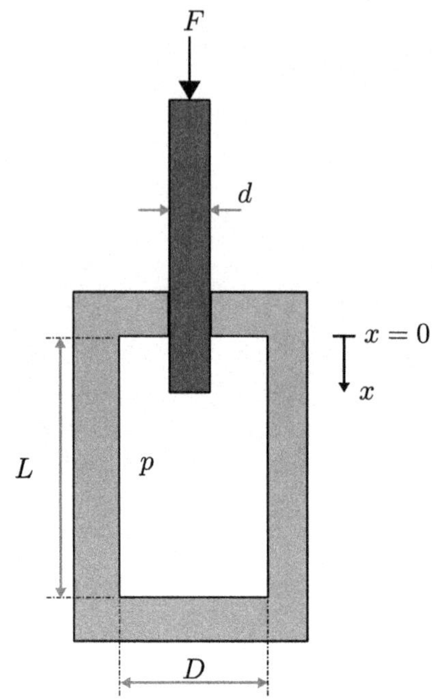

(c) At point two x_2 the force is $F_2 = 20F_1$. What is the pressure p_2 and what is the position x_2
(d) ? Derive a symbolic function $x(F)$, and sketch (x, F).

16.1.3 Viscous Force on Rotating Body

A shaft is rotating in a hydrostatic bearing as seen in Fig. 16.3. Determine the viscous friction force on the shaft when the width of the bearing is $L = 30$ mm, the outer and inner diameters are $D = 30$ mm and $d = 29.1$ mm respectively. The fluid is a Shell Tellus 32 with a temperature of 40 °C at which the kinematic viscosity is 32 cSt and the density is assumed to be $\rho = 860 \frac{\text{kg}}{\text{m}^3}$. Assume the pressure to be equal throughout the bearing. The rotational velocity of the shaft is $\omega = \frac{1000 \frac{\text{rev}}{\text{min}} 2\pi \frac{\text{rad}}{\text{Rev}}}{60 \frac{\text{s}}{\text{min}}}$ (Fig. 16.3).

Find;

(a) The torque on the shaft from the fluid, due to viscous forces.
(b) A function for the torque $\tau(d, D, \omega)$
(c) Discuss and investigate the effect of changing the clearance, $\delta = D - d$ when the rotational velocity is constant.

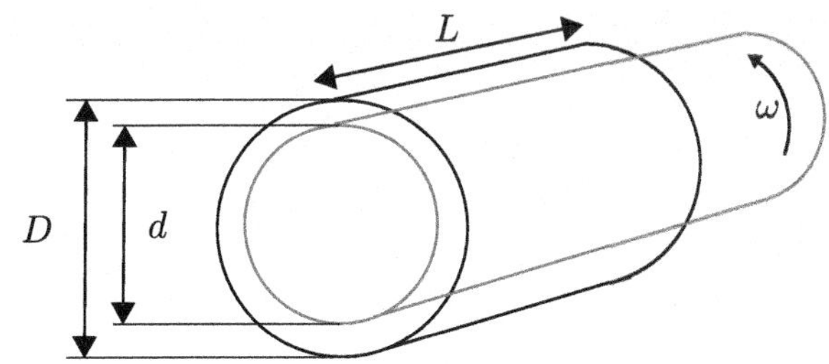

Fig. 16.3 Illustration for exercise Sect. 16.1.3

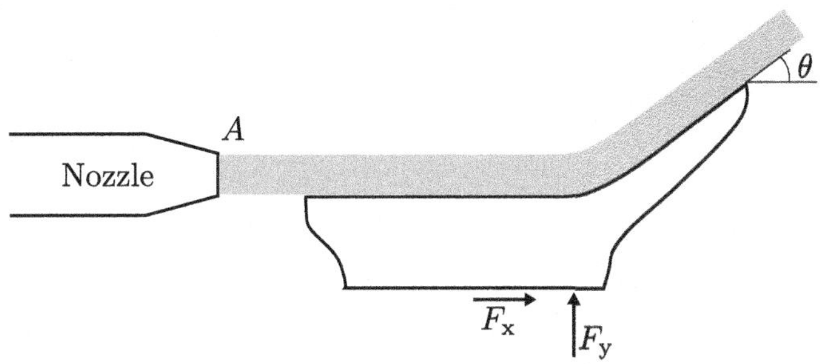

Fig. 16.4 Illustration for exercise Sect. 16.1.4

16.1.4 Fluid Momentum

A free fluid jet is as seen in Fig. 16.4 flowing over an engine vane. The fluid jet is leaving the nozzle in direction $n_1 = [1,\ 0,\ 0]$ and is due to the vane changing direction to $n_2 = [\cos\theta,\ \sin\theta,\ 0]$. Calculate the force required to hold the vane stationary, when gravity and viscous forces are neglected. (Note that the fluid jet is in free air such that the pressure is equal to atmospheric pressure and constant.) Let's use water as fluid and assume the density to be $\rho = 1000\ \frac{\text{kg}}{\text{m}^3}$.

(a) Make a choice of control volume.
(b) Use Reynolds Transport Theorem to form a function for the forces F_x and F_y.
(c) Calculate the size of the forces, assuming the flow area to be $20\ \text{mm}^2$ and constant along the flow; the fluid velocity is $5\ \frac{\text{m}}{\text{s}}$ and the angle θ is 30°.

16.2 Fluid Mechanics II

16.2.1 Orifice Flow I

Given the orifice in Fig. 16.5 with a discharge coefficient of 0.7 and an oil density of $886\ \frac{\text{kg}}{\text{m}^3}$.

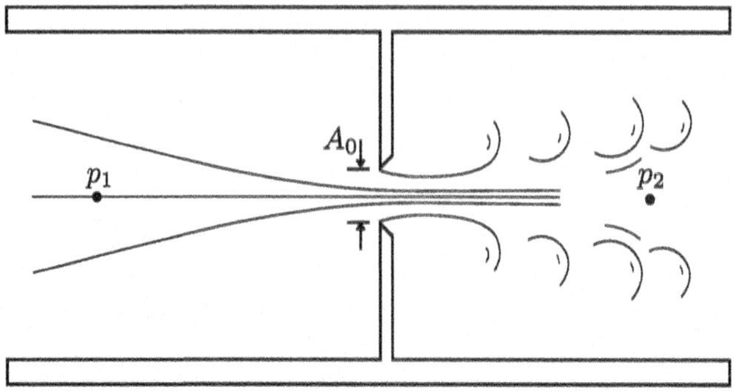

Fig. 16.5 Illustration for exercises Sect. 16.2.1

(a) Calculate the flow Q through the orifice with a flow area $A_0 = 10\ \text{mm}^2$ when the pressure drop from p_1 to p_2 is 100 bar.
(b) Determine Reynolds number (kinematic viscosity $\nu = 32$ cSt.) and comment on the result.

16.2.2 *Orifice Flow II*

Given the orifice in Fig. 16.5 with a discharge coefficient of 0.7 and an oil density of $886\ \frac{\text{kg}}{\text{m}^3}$.

(a) Calculate the area A_0 yielding a flow of $50\ \frac{\text{L}}{\text{min}}$ when the pressure drop from p_1 to p_2 is 100 bar.
(b) Determine Reynolds number (kinematic viscosity $\nu = 32$ cSt.) and comment on the result.

16.2.3 *Orifice Flow III*

The orifices in this exercise have equal size. $A_d = 10\ \text{mm}^2$ and an oil with density $886\ \frac{\text{kg}}{\text{m}^3}$ is used. The discharge coefficient is 0.7. The tank pressure is set to 0 bar.

Given the flow and pressure in the Fig. 16.6 calculate:

(a) The flow through the orifice.
(b) The flow through the orifices and the pressure p_2 between the orifices.
(c) The flow through the orifices and the pressure level after the pump.

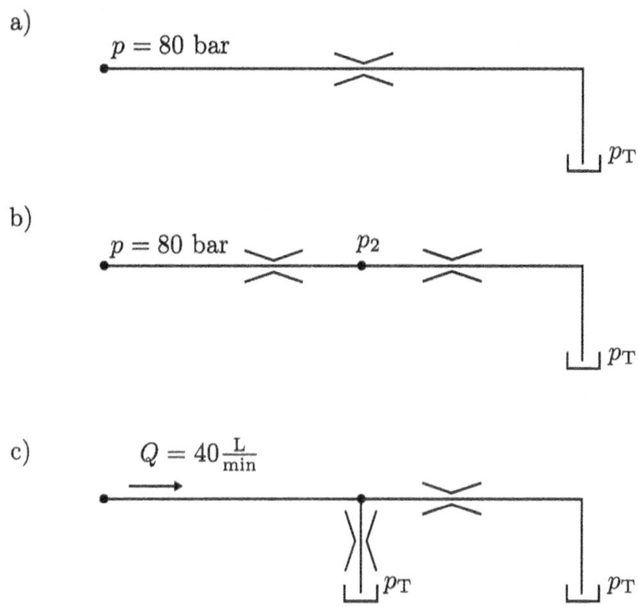

Fig. 16.6 Illustration for exercises Sect. 16.2.3

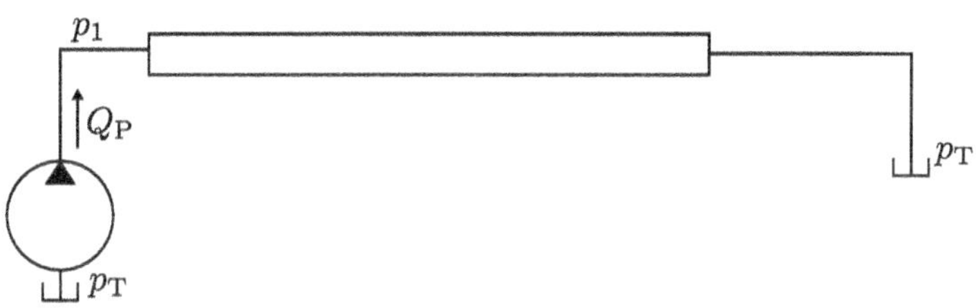

Fig. 16.7 Illustration for exercises Sect. 16.2.4

16.2.4 Pipe Flow I

A pump is supplying a flow Q_P which is led through a 20 m long pipe with a diameter of 10 mm and back to the tank (See Fig. 16.7). The fluid used has density and kinematic viscosity, $\rho = 860\ \text{kg/m}^3$ and $\nu = 32\frac{\text{mm}^2}{\text{s}}$. The tank pressure is set to 1 bar.

Calculate,

(a) the pressure p_1 for $Q_P = 10$ L/min
(b) the pressure p_1 for $Q_P = 20$ L/min
(c) the pressure p_1 for $Q_P = 40$ L/min.

16.2.5 Pipe Flow II

An oil flow of $Q = 30$ L/min is led through a 200 m long horizontal line with inner diameter of 20 mm. The pressure drop is 500000 Pa and the oil density is 866 kg/m^3.

What is the oil viscosity?

16.2.6 Pipe Flow III

The system in Fig. 16.8 consists of a fixed displacement pump running at constant speed, 1000 rpm, a 5 m long hose with a diameter of 12 mm and an orifice with opening and discharge coefficient of 10 mm^2 and 0.8 respectively. The oil used has a density of 900 kg/m ^{3} and a dynamic viscosity of $\mu = 32 \cdot 10^{-6} \cdot 900 \frac{\text{Ns}}{\text{m}^2}$. The tank pressure is set to 1 bar.

With a pump displacement of $D_\text{P} = 25 \frac{\text{cm}^3}{\text{rev}}$ and a volumetric efficiency the pump flow is 25 L/min.

(a) What is the flow through the orifice?
(b) What is the pressure p_2?
(c) What is the pressure p_1?
(d) What are the power losses in the orifice and hose respectively?

16.2.7 Velocity Profile in an Annular Flow

Derive the velocity flow profile for the fluid flow in an annular ring as illustrated in Fig. 16.9. Note that both walls have zero velocity and that no slip condition apply zero velocity of the fluid in contact with the walls.

Hint: use polar coordinates.

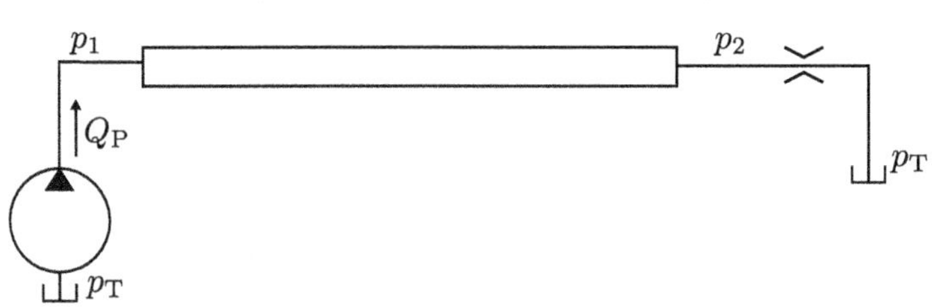

Fig. 16.8 Illustration for exercises Sect. 16.2.6

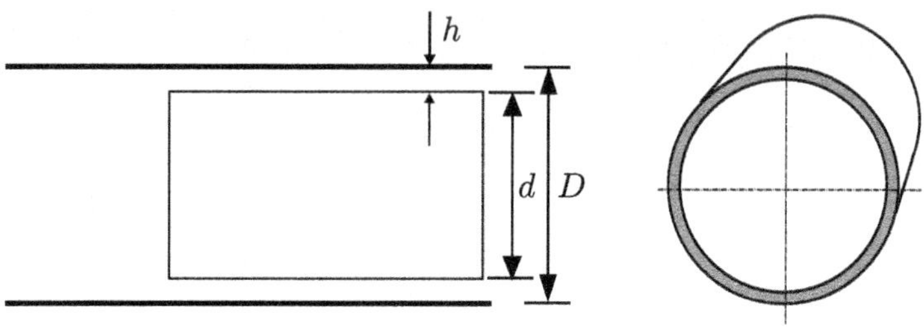

Fig. 16.9 Illustration for exercise Sect. 16.2.7. The grey ring indicates the flow cross section

16.3 Pumps, Motors and Cylinders

16.3.1 Pump I

A hydraulic pump driven by an electric motor is working at 1500 rpm against a pressure of 150 bar, hence the pressure at the pump outlet is 150 bar. The pump displacement is 50 cm^3. The pump flow is measured to 68 $\frac{L}{min}$ and the power on the output shaft of the electrical motor is measured to 20 kW. The suction pressure (pressure at the pump inlet) is approximately 0 bar. Under the given circumstances find:

(a) The volumetric efficiency, η_{Pv}, of the pump.
(b) The hydro-mechanical efficiency, η_{Phm}, of the pump.
(c) The total efficiency, η_P, of the pump.

16.3.2 Pump II

A hydraulic pump is working at 1000 rpm against a pressure of 100 bar delivering a flow of 20.8 $\frac{L}{min}$. Under these conditions it is known that the volumetric efficiency and total efficiency of the pump are 0.91 and 0.80 respectively. Knowing that the suction pressure is approximately 1 bar, determine:

(a) The hydro-mechanical efficiency, η_{Phm}, of the pump.
(b) The required input torque to drive the pump.

16.3.3 Motor I

A 25 cm^3 hydraulic motor is rotating a shaft at 1000 rpm delivering a torque of 35 Nm. The flow into the motor and the pressure drop across the motor are measured to 30 $\frac{L}{min}$ and 100 bar, respectively. Under the given circumstances, determine:

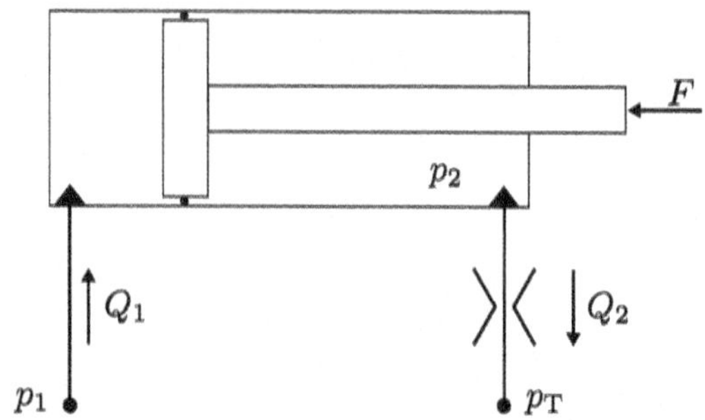

Fig. 16.10 Illustration for exercise Sect. 16.3.4

(a) The volumetric efficiency, η_{Mv} of the motor.
(b) The hydro-mechanical efficiency, η_{Mhm} of the motor.
(c) The total efficiency, η_M of the motor.

16.3.4 Cylinder I

A hydraulic differential cylinder with a piston diameter 40 mm and a piston rod diameter of 25 mm, see Fig. 16.10. The hydro-mechanical efficiency is 0.95. The sealing of the cylinder imposes a volumetric efficiency of 1. An orifice is connecting the rod (ring) side of the cylinder to tank pressure (0 bar), hence the pressure in the rod chamber is given as (16.1). With the orifice equation we yield:

$$p_2 = \frac{Q_2^2}{C_d^2 A_d^2 \sqrt{\frac{2}{\rho}}^{-2}} = \frac{Q_2^2}{0.7^2 \cdot (10 \cdot 10^{-6} \text{m}^2)^2 \sqrt{\frac{2}{860 \frac{\text{kg}}{\text{m}^3}}}^{-2}} \tag{16.1}$$

Calculate the required supply flow, Q_1, and pressure, p_1, to obtain a piston velocity of 0.4 $\frac{\text{m}}{\text{s}}$ when the load force is 20000N.

16.3.5 Cylinder II

A load of 2000 kg is to be lifted with a speed of 0.4 $\frac{\text{m}}{\text{s}}$. In Fig. 16.11 three ways to perform the lifting are suggested. All three cylinders are differential cylinders, 40/25/500 mm. The hydro-mechanical efficiency is 0.95 and the back pressure p_2 is 5 bar in load cases (a) and (b). The volumetric efficiency is 1. Use $g = 10\frac{\text{m}}{\text{s}^2}$.

In each of the three cases find:

1. The flow required to move the load at the given velocity, Q_1.
2. The required pressure, p_1.

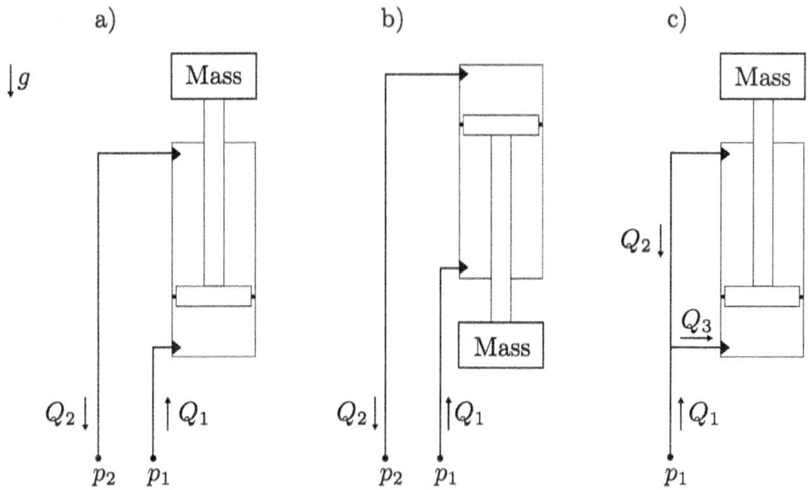

Fig. 16.11 Illustration for exercise Sect. 16.3.5

3. The power supplied, through the in flow.
4. The power deliver to the load.
5. Discuss the results.

16.3.6 Cylinder III

A fixed displacement pump is driven at 1500 RPM. The pump has a displacement of 20 cm^3 and is driving a differential cylinder 40/20/400 mm lifting a load of 1000 kg as shown in Fig. 16.12. The return line is connected to the tank through a fixed orifice with discharge coefficient and flow area of 0.7 and 10 mm^2 respectively. The hydro-mechanical efficiency of the cylinder is $\eta_{\mathrm{Chm}} = 0,95$, the volumetric efficiency is 1. The efficiencies for the pump are $\eta_{\mathrm{Phm}} = 0.95$ and $\eta_{\mathrm{Pv}} = 0.9$. The used oil has a density of 860 $\frac{\mathrm{kg}}{\mathrm{m}^3}$.

Calculate:

1. The flow delivered by the pump, Q_1.
2. The piston velocity.
3. The return flow, Q_2.
4. The cylinder chamber pressures, p_1 and p_2.
5. The power delivered to the load.
6. The hydraulic power delivered from the pump.
7. The required mechanical power to the pump.

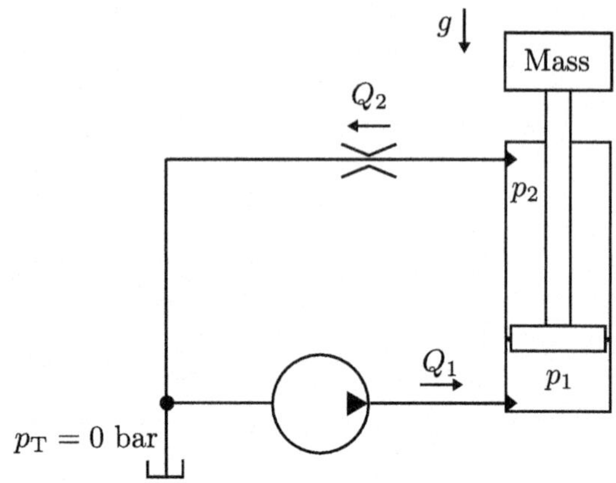

Fig. 16.12 Illustration for exercise Sect. 16.3.6

16.4 Steady State Analysis

16.4.1 System 1—Raising a Differential Cylinder

A differential cylinder with dimensions 40/25/400 mm is raising a load of 2000 kg. The fluid power system is in Fig. 16.13. Assume all components to be ideal such that all efficiencies are set to 1. The directional valves have opening areas $A_{PA} = A_{BT} = 18\ \text{mm}^2$ and a discharge coefficient of 0.7. The pressure relief valve is set to 250 bar. The fixed displacement pump has a displacement of 25 cm^3 and is driven by an electrical motor at a speed of 2000 rpm. Use a gravitational acceleration of 9.81 $\frac{\text{m}}{\text{s}^2}$ and an oil density of 900 $\frac{\text{kg}}{\text{m}^3}$ and calculate:

1. The piston velocity.
2. The flow through the directional valve, Q_{PA} and Q_{BT}.
3. The cylinder chamber pressure.
4. The pump pressure.

16.4.2 System 1—Lowering a Differential Cylinder

The load is now to be lowered (same system as in exercises Sect. 16.4.1). It is desired to have a pressure of 5 bar in the rod side chamber while lowering the load. A discharge coefficient of 0.7 is used for both PB and AT. The opening area from P to B is $A_{PB} = 18\ \text{m}^2$ while the A to T opening area is to be designed in sub question 5. Use the value from exercises Sect. 16.4.1 and calculate (Fig. 16.14):

1. The piston velocity and the flow, Q_{PB}, through the directional valve.
2. The flow, Q_{AT}, through the directional valve.
3. The pump pressure.

4. The cylinder chamber pressure.
5. The required orifice area, A_{AT}.

16.4.3 System 2—Flow Regeneration 1

A differential cylinder is supplied from a fixed displacement pump as seen in Fig. 16.15. The cylinder has dimensions: 50/40/1000 mm, a hydro-mechanical efficiency of 0.9 and volumetric efficiency of 1. The cylinder is moving a constant mass of 2500 kg. The pump has a fixed displacement of 30 $\frac{\text{cm}^3}{\text{rev}}$, and volumetric efficiency of 0.94 and hydro-mechanical efficiency of 0.92. The rotational velocity of the pump is 1000 rpm. The two cylinder chambers are in this configuration connected through

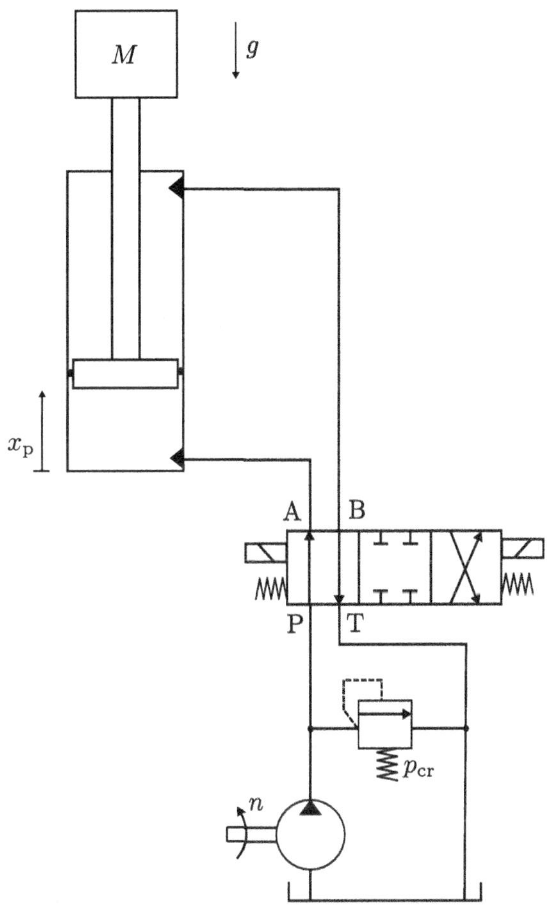

Fig. 16.13 Illustration for exercises Sect. 16.4.1

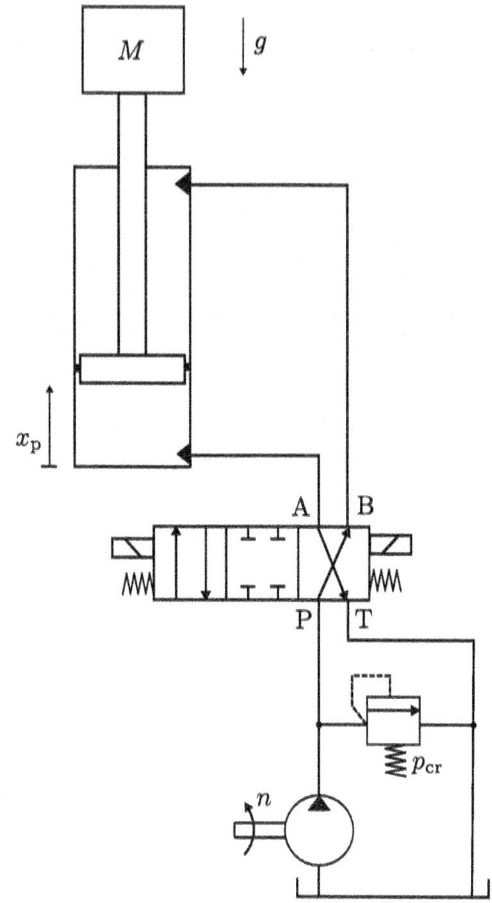

Fig. 16.14 Illustration for exercises Sect. 16.4.2

an orifice with an opening area of 10 mm^2 and a discharge coefficient of 0.6. The oil used has a density of 866 $\frac{\text{kg}}{\text{m}^3}$ and a kinematic viscosity of 32 $\frac{\text{mm}^2}{\text{s}}$.

1. Draw the flow directions on the diagram.
2. What is the flow out of the pump?
3. What is the steady state velocity for the piston?
4. What is the steady state pressure in the two cylinder chambers?
5. What will happen when the piston reaches end stop?

16.4.4 System 3—Flow Regeneration 2

A box with mass 1000 kg is moved by an asymmetric cylinder, see Fig. 16.16. The movement of the box on the rollers is frictionless. A constant load force of $F_L = 10$

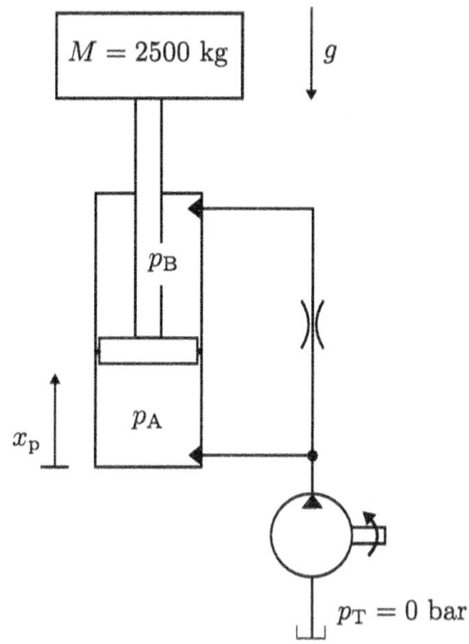

Fig. 16.15 Illustration for exercises Sect. 16.4.3

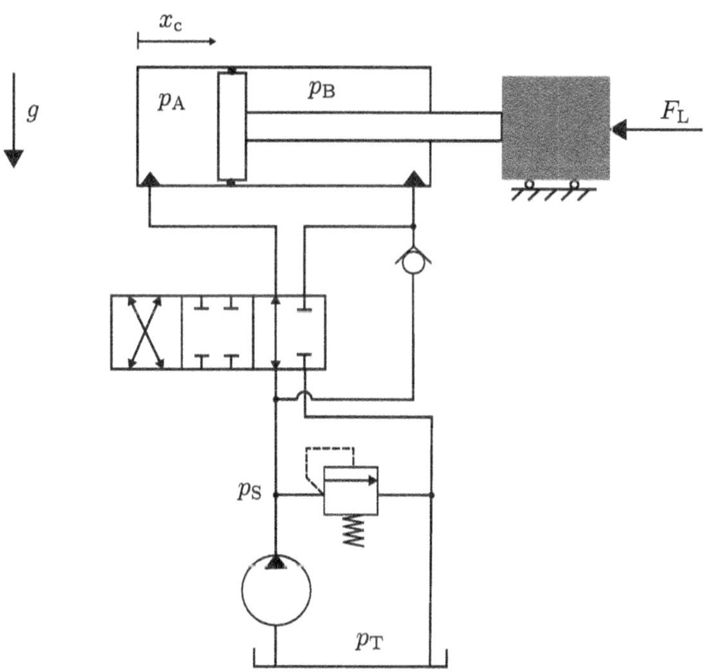

Fig. 16.16 Illustration for exercises Sect. 16.4.4

kN is seen. The cylinder has dimensions 50/40/600 mm and is controlled by a 4/3 directional valve in which all paths are equal with $A_0 = 20$ mm^2 and $C_d = 0.8$. The volumetric and hydro-mechanical efficiencies of the cylinder are both 1. A fixed displacement pump ($D_P = 30\frac{\text{cm}^3}{\text{rev}}$) rotating at 1500 rpm delivers fluid to the directional valve and is protected by an ideal pressure relief valve set to 250 bar. Both the pressure relief and the check valve are assumed ideal (no pressure drop). The volumetric and hydro-mechanical efficiencies of the pump are 0.9 and 0.85 respectively. Use an oil density of $\rho = 900\frac{\text{kg}}{\text{m}^3}$ and assume the tank pressure to be zero. With the configuration in the figure and steady state obtained:

(a) Draw flow direction of the diagram.
(b) What is the flow out of the pump?
(c) What is the piston velocity?
(d) What is the pressure at the pump outlet, p_S?
(e) What is the mechanical power delivered to the pump?
(f) What is the power loss in the directional valve?

16.4.5 System 4—Motor Lifting a Load

A load of 100 kg is moved with a wrench drive. The barrel has a radius of 300 mm. The hydraulic wrench motor is controlled with the directional valve V2. A fixed displacement pump is supplying fluid power to the system. Data for the system is given in Fig. 16.17. $\rho_{\text{oil}} = 900\frac{\text{kg}}{\text{m}^3}$.

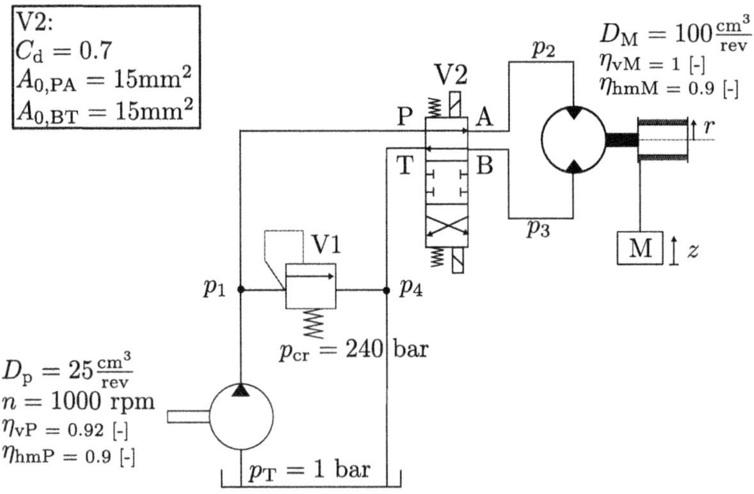

Fig. 16.17 Illustration for exercises Sect. 16.4.5

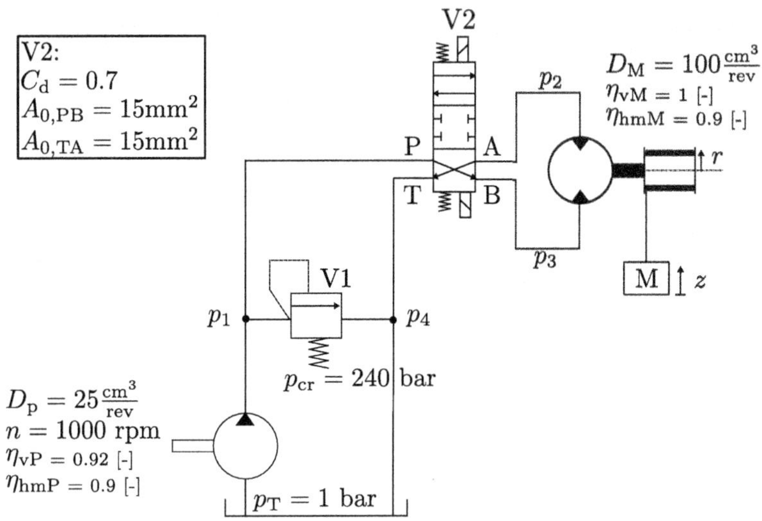

Fig. 16.18 Illustration for exercises Sect. 16.4.6

(a) Show that the mass is raised with a velocity of approximately ($\dot{z} = 7.23 \frac{\text{m}}{\text{s}}$).
(b) Determine the pressures in the system (p_4 to p_1).
(c) Determine the power requested on the pump shaft to drive it at the calculated operation.

16.4.6 System 4—Motor Lowering a Load

The system from exercise Sect. 16.4.5 is now lowering the load. As the pressure drop across the directional valve is small compared to the load pressure p_A, when supporting the load the current setup will not be able to lower the load controlled. You are asked to choose a valve type and size such that the load controlled may be lowered; the used directional valve must stay in the system (Fig. 16.18).

(a) Choose a valve type and placement.
(b) Choose settings of the valve, e.g. (A_0, Q, p_{cr})
(c) Perform a steady state analysis that validate your design.

16.5 System Modelling

16.5.1 System D1—Simple Pump Cylinder Drive

A fixed displacement pump controls a differential cylinder lifting a mass of 2500 kg against an external load, F_L. A 60x30x500 mm cylinder is used, see Fig. 16.19. The cylinder may be assumed ideal, such that leakage and friction in the cylinder are zero. The fixed displacement pump has a displacement of 30 cm^3 per revolution. The pump flow may be modelled as an ideal pump combined with laminar leakage flow Q_{le}, modelled using a leakage coefficient C_{le} = 0.05 L/min/bar. The pump is driven by a velocity controlled electric machine such that ω_p is the input velocity. The bulk modulus of the fluid is set constant to 10000 bar. The hose volume from the pump to the cylinder is 0.1 L. The B chamber is vented to air, such that p_B is equal p_T both

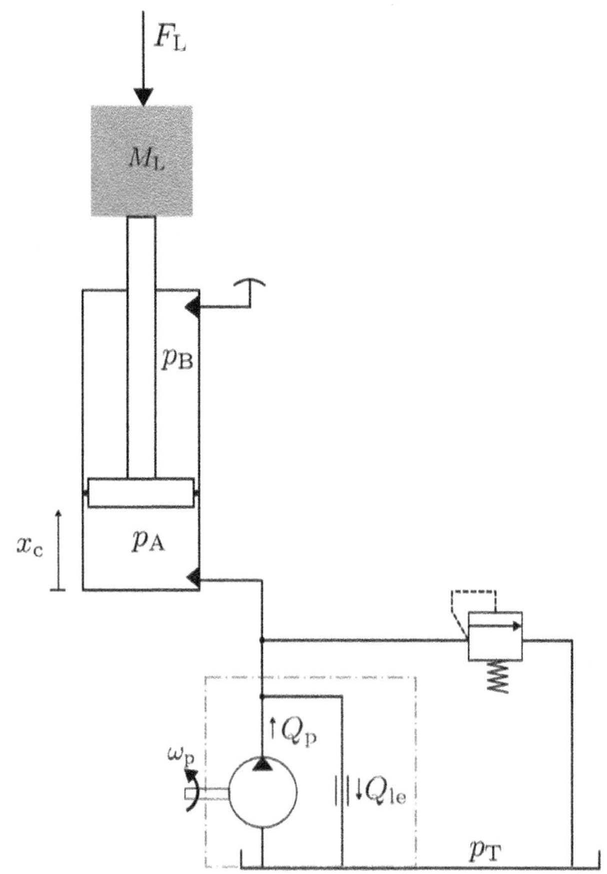

Fig. 16.19 Illustration for exercises Sect. 16.5.1

of which are set to 0 bar. The gravitational acceleration is 9.81 m/s^2. The pressure relief valve is assumed ideal and has an opening pressure of 250 bar.

1. Given a pump rotational velocity of 1000 rpm, what is the piston steady state velocity? With a load force, F_L, of;
    a. 10000 N
    b. 30000 N
    c. 50000 N
2. Set up the time domain model, with load force and pump velocity as inputs. (Assume the pressure relief valve to be closed.)

16.5.2 System D2—Hanging Mass

A 120 kg mass is moved vertically by a differential cylinder as seen in Fig. 16.20. It is a 40/25/400 mm cylinder, the pump is a fixed displacement pump running at

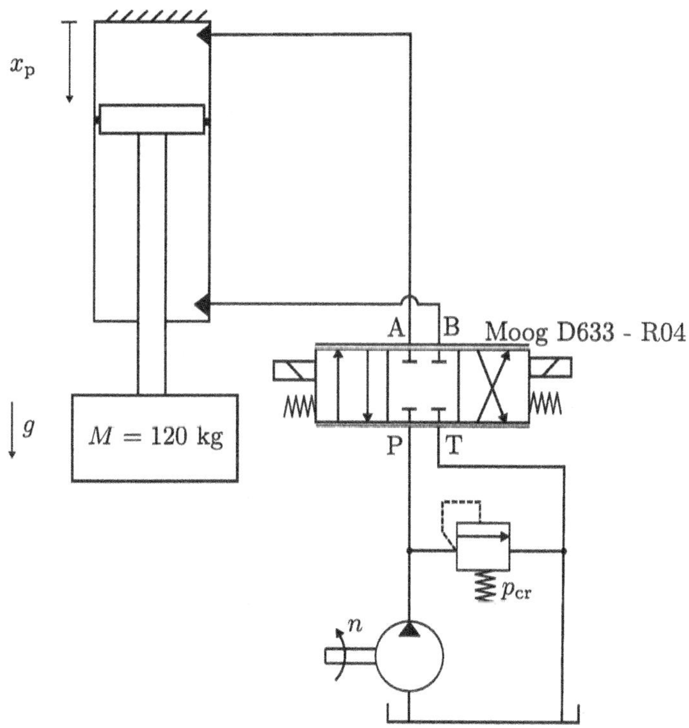

Fig. 16.20 Illustration for exercises Sect. 16.5.2

1500 rpm, the pressure relief valve is set to 120 bar. The proportional valve used is the Moog D633—R04 and it has a 10 L/min spool. The hose 1/2" connections from servo valve to cylinder chambers are approximately 1.5 m, likewise the 1/2" hose connection from pump and tank to the servo valve is approximately 1.5 m. The valve spool reference is system input.

1. Derive a time domain model for the system. (One may assume $p_S = p_{cr}$)
2. Implement the time domain model is MatLab SIMULINK
3. Discuss which measurements to perform in order that the simulation model can be validated.

16.6 Analysis of Dynamic Systems

16.6.1 Pilot Chamber I

A pilot chamber with fixed volume V_c is connected to a pressure source $p_S(t)$ through a fixed orifice, see Fig. 16.21. Assume a constant bulk modulus in the linearisation point.

(a) Derive the time domain flow equation for the flow into the chamber.
(b) Derive the time domain equation for the pressure gradient.
(c) Linearise the equations and Laplace transform them.
(d) Derive a transfer function with the chamber pressure $P_c(s)$ as output and the source pressure $P_S(s)$ as input.
(e) Comment on the results. What is the DC gain?

16.6.2 Pilot Chamber II

A pilot chamber with fixed volume V_c is connected to a pressure source $p_S(t)$ through a fixed orifice and constant tank pressure p_T through another fixed orifice, see Fig. 16.22. Assume a constant bulk modulus in the linearisation point.

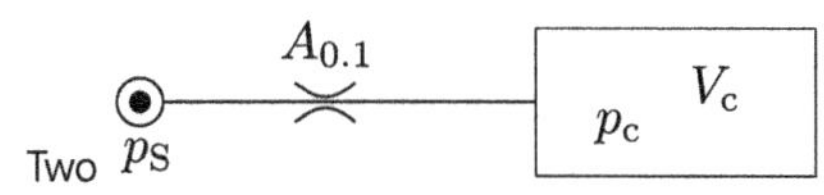

Fig. 16.21 Illustration for exercises Sect. 16.6.1

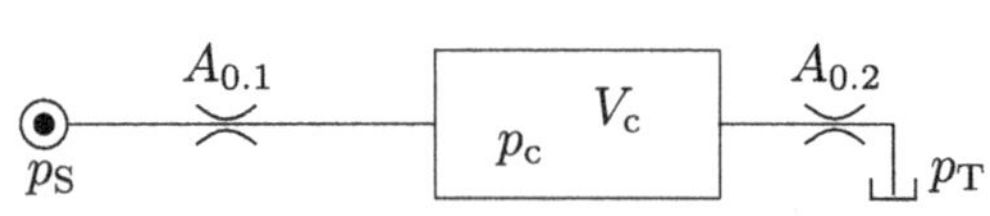

Fig. 16.22 Illustration for exercises Sect. 16.6.2

(a) Derive the time domain flow equation for the flow into the chamber.
(b) Derive the time domain flow equation for the flow out of the chamber.
(c) Derive the time domain equation for the chamber pressure gradient.
(d) Linearise the equations and Laplace transform them.
(e) Derive a transfer function with the chamber pressure $P_c(s)$ as output and the source pressure $P_S(s)$ as input assuming $p_T = 0$.
(f) Comment on the results. What is the DC gain?

16.6.3 Spring Loaded Accumulator

A spring loaded hydraulic cylinder is connected to a pressure source ($p_S = 100$ bar) through a variable orifice and to the tank through a fixed orifice, as illustrated in Fig. 16.23. The spring chamber is vented to air, such that the pressure is 0 bar. The tank pressure, p_T, is also set to 0 bar. The piston diameter is 10 mm and the spring constant is 10000 N/m. Both orifices have a discharge coefficient of 0.7. The fixed orifice has an opening area $A_{dT} = 2\text{mm}^2$ and the variable orifice has a linear opening between 0 mm^2 and 8 mm^2 ($A_{dS} = 8\ \text{mm}^2 \cdot x_v$, for $x_v = [0..1]$). The oil density is 860 $\frac{\text{kg}}{\text{m}^3}$ and the bulk modulus is assumed to be constant at 10000 bar. The piston has a mass of 10 kg and is subjected to viscous friction with a friction constant B = 100 Ns/m. When the pressure p_c is 0 bar the piston position is 0 m and the spring is unloaded.

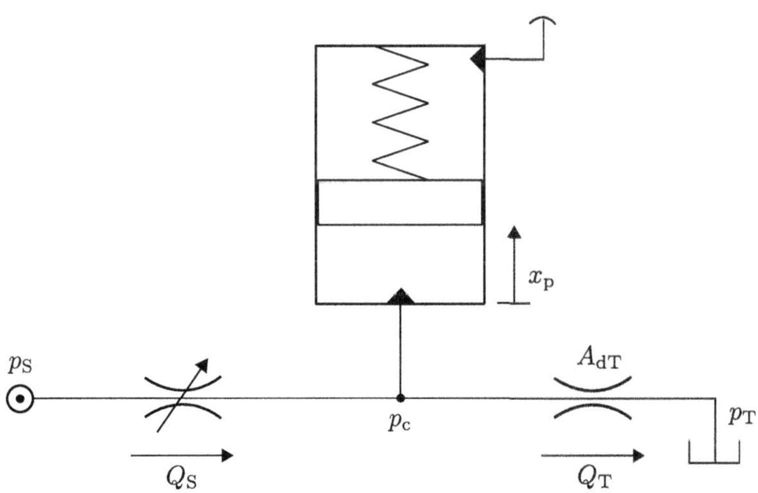

Fig. 16.23 Illustration for exercises Sect. 16.6.3

1. Show that the steady state piston pressure p_c is 50 bar when the variable valve is 25% open ($x_v = 0.25$).
2. What is the steady state piston position when the variable valve is 25% open ($x_v = 0.25$)?
3. Set up the dynamic equations for the system. Disregard the spring chamber, as the pressure is constant (0 bar).

Assume the chamber volume to be constant at the linearisation point and neglect the hose volumes.

4. Derive a linear model and construct the transfer function $\frac{X_p(s)}{X_V(s)}$.
5. Using $x_v = 0.25$ and the solutions from questions 1. and 2. as the linearisation point, what is the DC gain? Comment on the result.

16.6.4 Analysis of System D1

This exercise continues on from exercise Sect. 16.5.1, such that the time domain model from that exercise is the basis for the following analysis. Assume that the pump leakage coefficient is zero($C_{le} = 0$) in the next part and further assume the volume to be constant close to the linearisation point, such that $V_A(x_{c0})$ is used.

1. Derive a transfer function $\frac{sX(s)}{\omega p(s)}$, hence from pump velocity to cylinder velocity while assuming the load force to be constant, $F(s) = 0$;
2. What is the highest and lowest eigenfrequency for the system when linearised with a load force of zero? What are the corresponding linearisation points?
3. What is the analytical expression for the DC gain? Comment on the result.

16.6.5 Analysis of System D3—Hanging Mass

In this exercise, the system of exercise Sect. 16.5.2 is analysed and the model therein developed is used as the basis of this exercise.

1. Construct a linear model, symbolic (no numbers), for the system when assuming bulk modulus and chamber volumes to be constant in the linearisation point, volumes V_A and V_A is therefore not a function of piston position. (Exercise Sect. 16.5.2).
2. Choose a linearisation point such that the model is a Linear Time Invariant model.
3. Put the model on state space form.
4. Laplace transform the model and set up a block diagram.
5. Assuming supply and tank pressure are constant, set up a transfer function with cylinder piston velocity as output and valve spool position as input $\frac{sX_p(s)}{X_v(s)}$.

16.7 Control of Dynamic Systems

16.7.1 Velocity Control of System D3—Hanging Mass

Using the model constructed and the analysis performed for the Workshop System—Hanging Mass,

1. Design a classical P, PI, PD, and PID controller for velocity control.
2. Test the designed controllers in your time domain simulation model. Which reference velocity types (step, ramp, sinus) will you use?
3. Evaluate the controller structures and determine which are poor, ok and great for velocity control of the current system.

16.7.2 Position Control of System D3—Hanging Mass

Using the model constructed and the analysis performed for the Workshop System—Hanging Mass,

1. Design a classical P, PI, PD, and PID controller for position control.
2. Test the designed controllers in your time domain simulation model. Which reference velocity types (step, ramp, sinus) will you use?
3. Evaluate the controller structures and determine which are poor, ok and great for position control of the current system.

16.7.3 System Manipulation of System D3—Hanging Mass

Using the model constructed and the analysis performed for the Workshop System—Hanging Mass,

1. Design a pressure feedback loop to increase system damping.
2. Develop a flow feed forward structure that can be used together with cylinder piston velocity control.
3. Develop a flow feed forward structure that can be used together with cylinder piston position control.
4. Use an inverse orifice equation to change to a virtual flow input for the system.
5. Can any of the previously designed manipulation strategies be used together? Which and how?
6. Do valve dynamics influence the effect of the manipulation strategies?

Chapter 17
Solutions

Abstract This chapter holds solutions strategies for some exercises and may be for checking your own calculation of hints while working out the exercises.

17.1 Fluid Mechanics I

17.1.1 Fluid Compressibility

Using the continuity equation:

$$Q_{\text{in}} - Q_{\text{out}} = \dot{V} + \frac{V}{\beta}\dot{p}, \tag{17.1}$$

and setting the volume change and the inflow to zero, one gets:

$$Q_{\text{in}} = \frac{V}{\beta}\frac{dp}{dt}. \tag{17.2}$$

One may perform separation of variables and integrate:

$$\int_{t_0}^{t_f} Q_{\text{in}} dt = \int_{p(t_0)}^{p(t_f)} \tfrac{V}{\beta} dp. \tag{17.3}$$

As the volume and bulk modulus are constant these may be taken out of the integration, such that one gets:

$$\int_{t_0}^{t_f} Q_{\text{in}} dt = \tfrac{V}{\beta} \int_{p(t_0)}^{p(t_f)} 1 dp, \tag{17.4}$$

$$[Q_{\text{in}} t]_{t_0}^{t_f} = \tfrac{V}{\beta}(p(t_f) - p(t_0)). \tag{17.5}$$

From time t_0 to t_f the inflow will have accumulated to a volume; this volume we name V_β.

A. H. Hansen, *Fluid Power Systems*, Fluid Mechanics and Its Applications 129,
https://doi.org/10.1007/978-3-031-15089-0_17

$$V_\beta = \tfrac{V}{\beta}(p(t_f) - p(t_0)), \tag{17.6}$$
$$V_\beta = \tfrac{V}{\beta}\Delta p. \tag{17.7}$$

Note that the volume of fluid added is V_β and as the volume of the tank is not changed, the pressure must change. The amount of fluid that needs to be added to raise the pressure 100 bar is such:

$$V_\beta = \frac{0.100\text{m}^3}{10000 \cdot 10^5\,\text{Pa}} 100 \cdot 10^5\,\text{Pa} = 0.001\,\text{m}^3 = \underline{\underline{1\,\text{L}}}. \tag{17.8}$$

oOo

17.1.2 Fluid Spring

(a) The volume as function of rod position x is:

$$V(x) = D^2\frac{\pi}{4}L - d^2\frac{\pi}{4}x. \tag{17.9}$$

(b) The force at point 1 is given as:

$$F_1 = p_1 A = 10 \cdot 10^5\,\text{Pa}(0.030\,\text{m})^2\frac{\pi}{4} = \underline{\underline{706.86\,\text{N}}}. \tag{17.10}$$

(c) At point x_2 the force F_2 is 20 times F_1 and p_2 may as such be calculated by,

$$F_2 = p_2 A, \tag{17.11}$$
$$p_2 = \tfrac{F_2}{A} = \frac{14137\,\text{N}}{(0.030\,\text{m})^2\frac{\pi}{4}} = \underline{\underline{2.0 \cdot 10^7\,\text{Pa} \approx 200\,\text{bar}}}. \tag{17.12}$$

To calculated the position x_2 a function describing x_2 based on V_1, p_1, p_2 is derived from the definition of bulk modulus:

$$\beta = -V\tfrac{dp}{dV} \Rightarrow dp = -\tfrac{\beta}{V}dV, \tag{17.13}$$
$$\int 1dp = -\beta \int \tfrac{1}{V}dV, \quad \beta = \text{constant}, \tag{17.14}$$
$$[p]_{p_1}^{p_2} = -\beta \ln[V]_{V_1}^{V_2} \tag{17.15}$$

Note that the integration limits are set to the initial and final volume and pressure, such that:

$$p_2 - p_1 = -\beta[\ln(V_2) - \ln(V_1)] = \beta[\ln(V_1) - \ln(V_2)] = \beta \ln\left(\tfrac{V_1}{V_2}\right), \tag{17.16}$$

$$e^{\frac{p_2-p_1}{\beta}} = \tfrac{V_1}{V_2}, \tag{17.17}$$

$$V_2 = V_1 - Ax, \tag{17.18}$$

$$e^{\frac{p_2-p_1}{\beta}} = \tfrac{V_1}{V_1-Ax} \Rightarrow e^{\frac{p_2-p_1}{\beta}}(V_1 - Ax) = V_1 \tag{17.19}$$

$$(V_1 - Ax) = \frac{V_1}{e^{\frac{p_2-p_1}{\beta}}} \Rightarrow Ax = V_1 - \frac{V_1}{e^{\frac{p_2-p_1}{\beta}}} \tag{17.20}$$

The function for position therefore becomes:

$$x = \tfrac{V_1}{A}\left(1 - e^{\frac{p_1-p_2}{\beta}}\right), \tag{17.21}$$

$$x_2 = \frac{(0.1\,\text{m})^2 \frac{\pi}{4} \cdot 0.5\,\text{m}}{(0.03\,\text{m})^2 \frac{\pi}{4}} \left(1 - e^{\frac{10 \cdot 10^5\,\text{Pa} - 200 \cdot 10^5\,\text{Pa}}{15000 \cdot 10^5\,\text{Pa}}}\right) = \underline{\underline{0.0699\,\text{m}}}. \tag{17.22}$$

oOo

17.1.3 Viscous Force on Rotating Body

The viscous forces are shear forces, and are working on the surface area of the shaft in the bearing:

$$F = A_s \tau_s, \tag{17.23}$$

$$\tau_s = \mu \tfrac{d\dot{x}}{dr} = \mu \tfrac{\omega r_i}{r_o - r_i}, \tag{17.24}$$

$$A_s = d\pi L \tag{17.25}$$

The torque on the shaft depends on the shearing force and the radius of the shaft such that the torque is given as:

$$T = F r_i, \tag{17.26}$$

$$T = d\pi L \mu \frac{\omega r_i}{r_o - r_i} r_i. \tag{17.27}$$

Inserting the numbers one yields:

$$T = 0.0291\,\text{m}\pi 0.03\,\text{m} 0.0275\,\tfrac{\text{Ns}}{\text{m}^2} \tfrac{104.72\,\frac{1}{\text{s}} 0.0146\,\text{m}}{0.015\,\text{m} - 0.0146\,\text{m}} 0.0146 v\text{m} = \underline{\underline{0.0037\,\text{Nm}}} \tag{17.28}$$

A function for the torque depending on diameters and rotational velocity is derived to:

$$T(d, D, \omega) = d\pi L\mu \frac{\omega \frac{d}{2}}{\frac{D}{2}-\frac{d}{2}} \frac{d}{2} = \pi L\mu \frac{\frac{1}{4}\omega d^3}{\frac{1}{2}(D-d)}, \tag{17.29}$$

$$T(d, D, \omega) = \pi L\mu \frac{1}{2} \frac{\omega d^3}{D-d}. \tag{17.30}$$

oOo

17.1.4 Fluid Momentum

The control volume may be chosen as you like, however choosing it to be as in Fig. 17.1 yields the fluid flow to be perpendicular to the control volume surface when fluid flow crosses these.

Note that fluid is crossing the control volume surface at a 90° angle both at in- and outlet. We are to use RTT:

$$\frac{D}{Dt}(Mu) = \frac{\partial}{\partial t} \int_{cv} \rho u dV + \int_{cs} \rho u \bar{u} \cdot \hat{n}\, dA. \tag{17.31}$$

One may see that the problem is a planar problem in the xy-plane, hence a 2D vector equation is formed. Furthermore, one may note that the flow is steady, hence not changing in time and the time dependent first term is therefore zero and RTT reduces to:

$$\frac{D}{Dt}(Mu) = \int_{cs} \rho u \bar{u} \cdot \hat{n}\, dA. \tag{17.32}$$

Note that momentum is crossing the control surface in two places. The forces in the x and y directions are form RTT given as:

$$F_x = -Au_1^2\rho + Au_2^2\rho\cos\theta, \tag{17.33}$$

$$F_y = Au_2^2\rho\sin\theta. \tag{17.34}$$

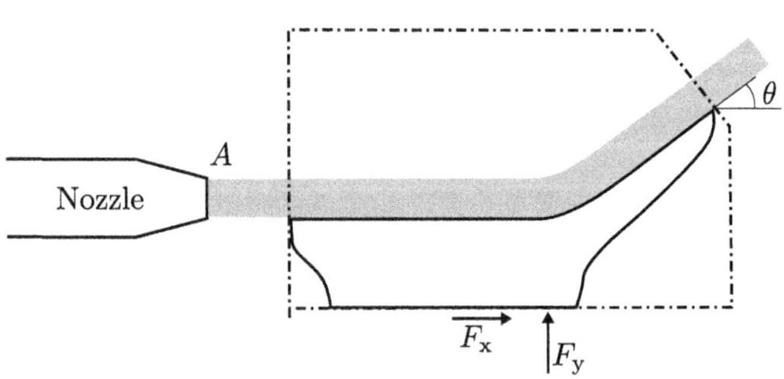

Fig. 17.1 Illustration for Exercise 16.1.4

Note that as the flow area is constant the speed of the fluid is constant; only the direction changes. Inserting the numbers one yields,

$$F_x = \underline{\underline{-0.067\ \text{N}}}, \quad (17.35)$$

$$F_y = \underline{\underline{0.25\ \text{N}}}. \quad (17.36)$$

oOo

17.2 Fluid Mechanics II

17.2.1 *Orifice Flow I*

(a) The orifice flow is found by:

$$Q = C_d A_0 \sqrt{\frac{2}{\rho}\Delta p}, \quad (17.37)$$

$$Q = 0.7 \cdot 10 \cdot 10^{-}6\,\text{m}^2 \sqrt{\frac{2}{886\,\frac{\text{kg}}{\text{m}^3}} 100 \cdot 10^5\,\text{Pa}} = \underline{\underline{0.0011\,\frac{\text{m}^3}{\text{s}}}}, \quad (17.38)$$

$$Q \approx \underline{\underline{63.1\,\frac{\text{L}}{\text{min}}}}. \quad (17.39)$$

(b) Reynolds number is determined as,

$$Re = \frac{\text{u}D_h}{\nu} = \frac{\frac{Q}{A}\sqrt{A\frac{4}{\pi}}}{\nu} \approx \underline{\underline{11727}} \quad (17.40)$$

Note that the diameter of the orifice area is used as hydraulic diameter.

oOo

17.2.2 *Orifice Flow II*

(a) The orifice area is found by:

$$A_0 = \frac{Q}{C_d\sqrt{\frac{2}{\rho}\Delta p}}, \quad (17.41)$$

$$A_0 = \frac{\frac{50\,\frac{\text{L}}{\text{min}}}{60\,\frac{\text{s}}{\text{min}}\,1000\,\frac{\text{L}}{\text{m}^3}}}{0.7\sqrt{\frac{2}{886\,\frac{\text{kg}}{\text{m}^3}} 100 \cdot 10^5\,\text{Pa}}} \approx \underline{\underline{7.9\ \text{mm}^2}}. \quad (17.42)$$

(b) Reynolds number is determined as,

$$Re = \frac{uD_h}{\nu} = \frac{\frac{Q}{A}\sqrt{A\frac{4}{\pi}}}{\nu} \approx \underline{\underline{10439}} \tag{17.43}$$

Note that the diameter of the orifice area is used as hydraulic diameter.

oOo

17.2.3 Orifice Flow III

(a) The flow through the orifice is found using the orifice equation,

$$Q = C_d A_d \sqrt{\frac{2}{\rho}\Delta p} \tag{17.44}$$

$$Q = 0.7 \cdot 10 \cdot 10^{-6}\,\text{m}^2 \sqrt{\frac{2}{886\,\frac{\text{kg}}{\text{m}^3}} 80 \cdot 10^5\ \text{Pa}} \tag{17.45}$$

$$= 0.00094\,\text{m}^3/\text{s} \approx \underline{\underline{56.44\,\frac{\text{L}}{\text{min}}}} \tag{17.46}$$

(b) The flow through the two orifices must be equal due to continuity:

$$Q_1 = Q_2, \tag{17.47}$$

$$Q_1 = C_d A_d \sqrt{\frac{2}{\rho}(p_1 - p_2)}, \tag{17.48}$$

$$Q_2 = C_d A_d \sqrt{\frac{2}{\rho}(p_2 - p_T)} \tag{17.49}$$

with the two orifices being equal, and equal flow running through the orifices, the pressure drop must be equal, $\Delta p = 40$ bar.

$$p = 80\ \text{bar} \Rightarrow p_2 = 40\ \text{bar} \tag{17.50}$$

$$Q = 0.7 \cdot 10 \cdot 10^{-6}\,\text{m}^2 \sqrt{\frac{2}{886\,\frac{\text{kg}}{\text{m}^3}} 40 \cdot 10^5\,\text{Pa}} \tag{17.51}$$

$$= 0.00067\,\frac{\text{m}^3}{\text{s}} \approx \underline{\underline{39.9\,\frac{\text{L}}{\text{min}}}} \tag{17.52}$$

(c) As the orifices are equal in size the pressure drop across the orifices are equal and the flow is distributed equally between the two orifices such that:

$$Q = Q_1 + Q_2, \tag{17.53}$$

$$Q_1 = C_d A_d \sqrt{\frac{2}{\rho}\Delta p}, \tag{17.54}$$

$$Q_2 = C_d A_d \sqrt{\frac{2}{\rho}\Delta p} \tag{17.55}$$

with the orifices and the pressure drop across them being equal, the flow through them is equal as well.

$$Q_1 = Q_2 \Rightarrow Q_1 = Q_2 = \frac{Q}{2} = \underline{\underline{20\,\frac{\text{L}}{\text{min}}}} \tag{17.56}$$

The pressure drop may now be determined as,

$$Q_1 = C_\text{d} A_\text{d} \sqrt{\frac{2}{\rho} \Delta p}, \tag{17.57}$$

$$\Delta p = \frac{Q_1^2}{\left(C_\text{d} A_\text{d} \sqrt{\frac{2}{\rho}}\right)^2}, \tag{17.58}$$

$$\Delta p = \frac{\left(\frac{20}{60000}\,\frac{\text{m}^3}{\text{s}}\right)^2}{\left(0.7 \cdot 10 \cdot 10^{-6}\,\text{m}^2 \sqrt{\frac{2}{886}\,\frac{\text{kg}}{\text{m}^3}}\right)^2} = 10.05 \cdot 10^5\ \text{Pa}, \tag{17.59}$$

$$p = \underline{\underline{10\ \text{bar}}}. \tag{17.60}$$

oOo

17.2.4 Pipe Flow I

The pressure drop may be calculated with:

$$\Delta p = \lambda \frac{L_\text{h}}{d_\text{h}} \rho \frac{\mathbf{u}^2}{2}, \tag{17.61}$$

where the average flow velocity is here:

$$\mathbf{u} = \frac{Q}{A_\text{h}} = \frac{Q}{d_\text{h}^2 \frac{\pi}{4}}, \tag{17.62}$$

and the friction coefficient *lambda* is given as:

$$\lambda = \frac{64}{Re}, \quad \text{laminar flow} \tag{17.63}$$

$$\lambda = 0.3164 \frac{1}{Re^{\frac{1}{4}}}, \quad \text{turbulent flow.} \tag{17.64}$$

Flow 10 L/min:

$$\Delta p = \frac{64}{\text{Re}_1} \frac{20\text{m}}{0.010\text{m}} 860 \frac{\text{kg}}{\text{m}^3} \frac{\left(\frac{10/60000\,\text{m}^3/s}{(0.010\,\text{m})^2 \pi/4}\right)^2}{2} = 3.7\ \text{bar}, \tag{17.65}$$

$$\text{Re}_1 = \frac{Q_1 d_\text{h}}{A_\text{h} \nu} = \frac{10/60000\ \text{m}^3/\text{s} \cdot 0.010\,\text{m}}{(0.010\,\text{m})^2 \pi/4 \cdot 32 \cdot 10^{-6}\,\frac{m^2}{s}} \approx 663 \tag{17.66}$$

Flow 20 L/min:

$$\Delta p = \frac{64}{\text{Re}_2}\frac{20\,\text{m}}{0.010\,\text{m}}860\,\frac{\text{kg}}{\text{m}^3}\frac{\left(\frac{20/60000\,\text{m}^3/\text{s}}{(0.010\,\text{m})^2\pi/4}\right)^2}{2} = 7.5\,\text{bar}, \tag{17.67}$$

$$\text{Re}_2 = \frac{Q_2 d_\text{h}}{A_\text{h}\nu} = \frac{20/60000\ \text{m}^3/\text{s}\cdot 0.010\,\text{m}}{(0.010\,\text{m})^2\pi/4\cdot 32\cdot 10^{-6}\,\frac{m^2}{s}} \approx 1326 \tag{17.68}$$

Flow 40 L/min:

$$\Delta p = 0.3164\frac{1}{\text{Re}_3^{\frac{1}{4}}}\frac{20\,\text{m}}{0.010\,\text{m}}860\,\frac{\text{kg}}{\text{m}^3}\frac{\left(\frac{40/60000\,\text{m}^3/\text{s}}{(0.010\text{m})^2\pi/4}\right)^2}{2} = 27.38\,\text{bar}, \tag{17.69}$$

$$\text{Re}_3 = \frac{Q_3 d_\text{h}}{A_\text{h}\nu} = \frac{40/60000\ \text{m}^3/\text{s}\cdot 0.010\,\text{m}}{(0.010\,\text{m})^2\pi/4\cdot 32\cdot 10^{-6}\,\frac{m^2}{s}} \approx 2652, \quad \text{Turbulent flow.} \tag{17.70}$$

The pressure p_1 will be [4.7 8.5 28.4] bar.

oOo

17.2.5 *Pipe Flow II*

The flow is assumed to be laminar such that we may use Darcy's equation:

$$Q = \frac{d^2\pi}{128\mu L}\Delta p, \tag{17.71}$$

$$\mu = \frac{\Delta p\pi d_\text{h}^4}{128 Q L_\text{h}}, \tag{17.72}$$

$$\mu = \frac{500000\,\text{Pa}\pi(0.02m)^4}{128\frac{30}{60000}\,\frac{\text{m}^3}{\text{s}}\,200\,\text{m}} = \underline{\underline{0.0196\,\frac{\text{Ns}}{\text{m}^s}}} \tag{17.73}$$

Lastly, we check that the flow is indeed laminar:

$$\text{Re} = \frac{\rho Q d_\text{h}}{A_\text{h}\mu} = 1404 \tag{17.74}$$

With this Reynolds number we expect a laminar flow.

oOo

17.2.6 *Pipe Flow III*

The flow out of the pump is 25 L/min.

(a) With the pump flow being 25 L/min the flow through both the pipe and the orifice must be 25 L/min, as the flow is steady and conservation of mass leads to flow continuity.
(b) The pressure drop across the orifice is found using the orifice equation:

$$Q = C_{\mathrm{d}} A_0 \sqrt{\tfrac{2}{\rho}\Delta p}, \tag{17.75}$$

$$\Delta p = \left(\frac{Q}{C_{\mathrm{d}} A_0 \sqrt{\frac{2}{\rho}}}\right)^2 = \left(\frac{25/60000\,\frac{\mathrm{m}^3}{\mathrm{s}}}{0.8\cdot 10\cdot 10^{-6}\,\mathrm{m}^2\sqrt{\frac{2}{900\,\frac{\mathrm{kg}}{\mathrm{m}^3}}}}\right)^2 = 12.2\,\mathrm{bar} \tag{17.76}$$

The pressure p_2 is 13.2 bar.

(c) We firstly calculate Reynolds number to check the flow regime:

$$\mathrm{Re} = \frac{\rho Q d_{\mathrm{h}}}{A_{\mathrm{h}}\mu} \approx 1382, \tag{17.77}$$

which indicates a laminar flow regime such that we may find the pressure drop through the pipe to be:

$$\Delta p = \frac{Q 128 \mu L_{\mathrm{h}}}{d_{\mathrm{h}}^4 \pi} = \frac{25/60000\,\frac{\mathrm{m}^3}{\mathrm{s}}\, 12832 \cdot 10^{-6} \cdot 900\,\frac{\mathrm{Ns}}{\mathrm{m}^2}\, 5\,\mathrm{m}}{(0.012\,\mathrm{m})^4 \pi} \approx 1.18\,\mathrm{bar} \tag{17.78}$$

The pressure p_1 is $13.2 + 1.18 = 14.38$ bar.

(d) The power loss across the orifice and the pipe is the pressure drop multiplied by the flow:

$$\mathcal{P}_{\mathrm{orifice}} = 12.2 \cdot 10^5\,\mathrm{Pa} 25/60000\,\frac{\mathrm{m}^3}{\mathrm{s}} = 508.6\,\mathrm{W}, \tag{17.79}$$

$$\mathcal{P}_{\mathrm{hose}} = 1.18 \cdot 10^5\,\mathrm{Pa} 25/60000\,\frac{\mathrm{m}^3}{\mathrm{s}} = 49.1\,\mathrm{W}. \tag{17.80}$$

oOo

17.2.7 *Velocity Profile in an Annular Flow*

Navier-Stokes in the $z-$direction assuming no flow velocity in the r and θ directions.

$$\rho\left(\frac{\partial u_z}{\partial t}\right) = -\frac{\partial p}{\partial z} + \mu\left(\frac{1}{r}\frac{\partial}{\partial r}\left(r\frac{\partial u_z}{\partial r}\right)\right). \tag{17.81}$$

If one assumes the flow to be steady, $\frac{\partial u_z}{\partial t} = 0$ we yield:

$$0 = -\frac{\partial p}{\partial z} + \mu\left(\frac{1}{r}\frac{\partial}{\partial r}\left(r\frac{\partial u_z}{\partial r}\right)\right). \tag{17.82}$$

Integration yields,

$$u_z = \frac{1}{4\mu}\frac{\partial p}{\partial z} r^2 + C_1 \ln r + C_2, \tag{17.83}$$

which is the same as when deriving the pipe flow; the variation is in the velocity limits at the walls. No-slip conditions at the two walls,

$$u_z(r_i) = 0, \quad (17.84)$$
$$u_z(r_o) = 0, \quad (17.85)$$

which is used to calculate the integration constants C_1 and C_2:

$$u_z(r_i) = 0 = \frac{1}{4\mu}\frac{\partial p}{\partial z}r_i^2 + C_1 \ln r_i + C_2, \quad (17.86)$$
$$u_z(r_o) = 0 = \frac{1}{4\mu}\frac{\partial p}{\partial z}r_o^2 + C_1 \ln r_o + C_2. \quad (17.87)$$

We start with C_1 yielding:

$$-C_2 = \frac{1}{4\mu}\frac{\partial p}{\partial z}r_i^2 + C_1 \ln r_i, \quad (17.88)$$
$$0 = \frac{1}{4\mu}\frac{\partial p}{\partial z}r_o^2 + C_1 \ln r_o - \frac{1}{4\mu}\frac{\partial p}{\partial z}r_i^2 - C_1 \ln r_i, \quad (17.89)$$
$$C_1(\ln r_o - \ln r_i) = \frac{1}{4\mu}\frac{\partial p}{\partial z}\left(r_i^2 - r_o^2\right), \quad (17.90)$$
$$C_1 = \frac{1}{4\mu}\frac{\partial p}{\partial z}\frac{\left(r_i^2 - r_o^2\right)}{\frac{\ln r_o}{\ln r_i}}. \quad (17.91)$$

Now C_2 is determined by:

$$0 = \frac{1}{4\mu}\frac{\partial p}{\partial z}r_o^2 + C_1 \ln r_o + C_2, \quad (17.92)$$
$$C_2 = -\frac{1}{4\mu}\frac{\partial p}{\partial z}r_o^2 + C_1 \ln r_o, \quad (17.93)$$
$$C_2 = -\frac{1}{4\mu}\frac{\partial p}{\partial z}r_o^2 + \frac{1}{4\mu}\frac{\partial p}{\partial z}\frac{\left(r_i^2 - r_o^2\right)}{\frac{\ln r_o}{\ln r_i}}\ln r_o, \quad (17.94)$$
$$C_2 = -\frac{1}{4\mu}\frac{\partial p}{\partial z}\left(r_o^2 + \frac{\left(r_i^2 - r_o^2\right)}{\frac{\ln r_o}{\ln r_i}}\ln r_o\right). \quad (17.95)$$

Inserting C_1 and C_2 yields:

$$u_z = \frac{1}{4\mu}\frac{\partial p}{\partial z}r^2 + \frac{1}{4\mu}\frac{\partial p}{\partial z}\frac{\left(r_i^2 - r_o^2\right)}{\frac{\ln r_o}{\ln r_i}}\ln r - \frac{1}{4\mu}\frac{\partial p}{\partial z}\left(r_o^2 + \frac{\left(r_i^2 - r_o^2\right)}{\frac{\ln r_o}{\ln r_i}}\ln r_o\right). \quad (17.96)$$

The velocity profile function then becomes:

$$u_z(r) = \frac{1}{4\mu}\frac{\partial p}{\partial z}\left(r^2 - r_o^2 + \frac{r_i^2 - r_o^2}{\frac{\ln r_o}{\ln r_i}}\ln\frac{r}{r_o}\right). \quad (17.97)$$

The fluid flow is given as;

$$Q = \int_{r_i}^{r_o} u_z(r)\pi 2r dr \quad (17.98)$$

oOo

17.3 Pumps, Motors and Cylinders

17.3.1 *Pump I*

(a) The volumetric efficiency, η_{Pv}, of the pump is calculated from the actual measured pump flow and the theoretical pump flow:

$$\eta_{vP} = \frac{Q_P}{Q_{tP}} = \frac{68\,\frac{L}{min}}{50\,\frac{cm^3}{rev}\cdot 1500\,\frac{rev}{min}\,\frac{1L}{1000\,cm^3}} = \underline{\underline{0.91}} \tag{17.99}$$

(b) The hydro-mechanical efficiency, η_{Phm}, of the pump is calculated as the ratio of theoretical torque to drive the pump and actual torque delivered from the electrical motor:

$$\eta_{hmP} = \frac{\tau_{tP}}{\tau_P} = \frac{\Delta p_P D_\omega}{\frac{\mathcal{P}_{El}}{\omega}} = \underline{\underline{0.94}} \tag{17.100}$$

(c) The total efficiency, η_P, of the pump is the volumetric and the hydro-mechanical efficiencies multiplied:

$$\eta_P = \eta_{vP}\eta_{hmP} = \underline{\underline{0.85}}, \tag{17.101}$$

or the hydraulic pump delivered by the pump relative to the power delivered from the electrical motor:

$$\eta_P = \frac{Q_P \Delta p_P}{\mathcal{P}_{El}} = \frac{68\,\frac{L}{min}\,\frac{1\,min}{60s}\,\frac{1\,m^3}{1000\,L}\,150\cdot 10^5\,Pa}{20000\,W} = \underline{\underline{0.85}}. \tag{17.102}$$

oOo

17.3.2 *Pump II*

(a) The hydro-mechanical efficiency, η_{Phm}, of the pump is found from the total pump efficiency and the volumetric pump efficiency:

$$\eta_{hmP} = \frac{\eta_P}{\eta_{vP}} = \frac{0.80}{0.91} = \underline{\underline{0.88}} \tag{17.103}$$

(b) The required input torque to drive the pump may be found from the power delivered by the prime mover:

$$\mathcal{P}_{\text{hyd}} = Q_{\text{P}}(p_{\text{S}} - p_{\text{T}}) = 25.8\,\frac{\text{L}}{\text{min}}(100 - 1)10^5\,\text{Pa} = 3432\,\text{W}, \tag{17.104}$$

$$\mathcal{P}_{\text{Motor}} = \frac{Q_{\text{P}}}{\eta_{\text{P}}} = \frac{3432\,\text{W}}{0.8} = 4290\text{W}, \tag{17.105}$$

$$\tau_{\text{Motor}} = \frac{\mathcal{P}_{\text{Motor}}}{\omega_{\text{P}}} = \frac{4290\,\text{W}}{\frac{1000\cdot 2\pi}{60}\,\frac{1}{s}} = \underline{\underline{41 Nm}}. \tag{17.106}$$

Another approach is to calculate the pump displacement and use that together with the pressure increase and hydro-mechanical efficiency:

$$D_{\text{P}} = \frac{Q_{\text{tP}}}{n} = \frac{Q_{\text{P}}}{n\eta_{\text{vP}}} = \frac{20.8\,\frac{\text{L}}{\text{min}}}{0.91\cdot 1000\,\frac{\text{rev}}{\text{min}}} \cdot 1000\,\frac{\text{cm}^3}{\text{L}} = 272.85\,\frac{\text{cm}^3}{\text{rev}}. \tag{17.107}$$

The required torque is such that:

$$\tau_{\text{P}} = \frac{\tau_{\text{tP}}}{\eta_{\text{hmP}}} = \frac{D_{\omega}\Delta p}{\eta_{\text{hmP}}} = \underline{\underline{41\ \text{Nm}}}. \tag{17.108}$$

oOo

17.3.3 Motor I

(a) The volumetric efficiency, η_{Mv} is calculated as:

$$\eta_{\text{vM}} = \frac{Q_{\text{tM}}}{Q_{\text{M}}} = \frac{25\,\frac{\text{cm}^3}{\text{rev}}\,1000\,\frac{\text{rev}}{\text{min}}\,\frac{1\text{L}}{1000\,\text{cm}^3}}{30\,\frac{\text{L}}{\text{min}}} = \underline{\underline{0.83}}. \tag{17.109}$$

(b) The hydro-mechanical efficiency, η_{Mhm}

$$\eta_{\text{hmM}} = \frac{\tau_{\text{M}}}{\tau_{\text{tM}}} = \frac{35\,\text{Nm}}{100\cdot 10^5\,\text{Pa}\frac{25\cdot 10^{-6}\,\text{m}^3}{2\pi}} = \underline{\underline{0.88}}. \tag{17.110}$$

(c) The total efficiency, η_{M}

$$\eta_{\text{M}} = \eta_{\text{vM}}\eta_{\text{hmM}} = \underline{\underline{0.73}}. \tag{17.111}$$

oOo

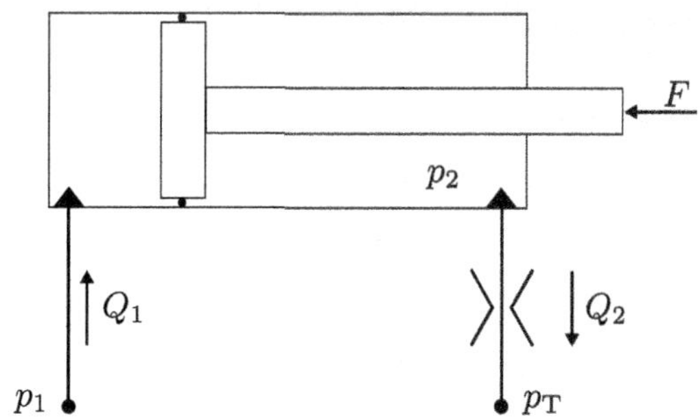

Fig. 17.2 Illustration for Exercise 16.3.4

17.3.4 Cylinder I

Assuming steady state and using the continuity equation, in- and outflow are calculated for a velocity of $0.4\,\frac{\text{m}}{\text{s}}$ (Fig. 17.2):

$$Q_1 = \dot{x}_p A_1 = \dot{x}_p D^2 \frac{\pi}{4} = 0.4\,\frac{\text{m}}{\text{s}} (0.04\,\text{m})^2 \frac{\pi}{4} = 30.16\,\frac{\text{L}}{\text{min}}. \tag{17.112}$$

$$Q_2 = \dot{x}_p A_2 = \dot{x}_p (D^2 - d^2)\tfrac{\pi}{4}, \tag{17.113}$$

$$Q_2 = 0.4\,\tfrac{\text{m}}{\text{s}}((0.065\,\text{m})^2 - (0.010\,\text{m})^2)\tfrac{\pi}{4} = \underline{\underline{18.38\,\tfrac{\text{L}}{\text{min}}}}. \tag{17.114}$$

One may note that the ratio between the in- and outflow matches the piston and rod side area. The back pressure is:

$$p_2 = \frac{\left(\frac{18.38}{60000}\,\frac{\text{m}^3}{\text{s}}\right)^2}{0.7^2 \cdot (30 \cdot 10^{-6}\,\text{m}^2)^2 \frac{2}{860\,\frac{\text{kg}}{\text{m}^3}}} \approx \underline{\underline{8.2\ \text{bar}}} \tag{17.115}$$

Using Newton's Second Law the pressure p_1 is calculated:

$$0 = \eta_{\text{hmC}}\,(p_1 A_1 - p_2 A_2) - 20000\,\text{N}, \tag{17.116}$$

$$p_1 = \frac{\frac{20000\,\text{N}}{\eta_{\text{hmC}}} + p_2 A_2}{A_1} = \underline{\underline{172.6\ \text{bar}}}. \tag{17.117}$$

oOo

Table 17.1 Results for Exercises 16.3.5

	(a)	(b)	(c)
Q_1 [L/min]	30.16	18.38	11.78
p_1 [bar]	170.6	283.1	428.9
$\mathcal{P}_{Q1}$ [kW]	8.57	8.67	8.42
$\mathcal{P}_{\text{Load}}$ [kW]	8	8	8

17.3.5 Cylinder II

Initially the results are shown in Table 17.1 and later the equations employed are given. Note that the power delivered in the three cases are close to equal, however, for significantly varying flow and pressure levels.

- Flow and pressure equations for system a,

$$Q_1 = D^2\tfrac{\pi}{4}\dot{x}_p, \tag{17.118}$$
$$Q_2 = (D^2 - d^2)\tfrac{\pi}{4}\dot{x}_p, \tag{17.119}$$
$$0 = \eta_{\text{hmP}}\left(p_1 D^2\tfrac{\pi}{4} - p_2(D^2 - d^2)\tfrac{\pi}{4}\right) - F_{\text{load}}. \tag{17.120}$$

- Flow and pressure equations for system b,

$$Q_1 = (D^2 - d^2)\tfrac{\pi}{4}\dot{x}_p, \tag{17.121}$$
$$Q_2 = D^2\tfrac{\pi}{4}\dot{x}_p, \tag{17.122}$$
$$0 = \eta_{\text{hmP}}\left(p_1(D^2 - d^2)\tfrac{\pi}{4} - p_2 D^2\tfrac{\pi}{4}\right) - F_{\text{load}}. \tag{17.123}$$

- Flow and pressure equations for system c,

$$0 = Q_1 + Q_2 - Q_3, \tag{17.124}$$
$$Q_1 = -(D^2 - d^2)\tfrac{\pi}{4}\dot{x}_p + D^2\tfrac{\pi}{4}\dot{x}_p \tag{17.125}$$
$$Q_1 = d^2\dot{x}_p, \tag{17.126}$$
$$0 = \eta_{\text{hmP}}\left(p_1 D^2\tfrac{\pi}{4} - p_1(D^2 - d^2)\tfrac{\pi}{4}\right) - F_{\text{load}}, \tag{17.127}$$
$$0 = \eta_{\text{hmP}}\left(p_1 d^2\tfrac{\pi}{4}\right) - F_{\text{load}}. \tag{17.128}$$

- The power equations are the same but with various inputs,

$$\mathcal{P}_{\text{Load}} = \dot{x}_p F_{\text{Load}}, \tag{17.129}$$
$$\mathcal{P}_{Q1} = Q_1 p_1. \tag{17.130}$$

oOo

17.3.6 Cylinder III

The equations are set up in the order they need to be solved, however all results are shown at the end.

Calculate:

1. The flow delivered by the pump, Q_1

$$Q_1 = \eta_{vP} D_p n \tag{17.131}$$

2. The piston velocity

$$\dot{x}_p = \frac{Q_1}{D^2 \frac{\pi}{4}} \tag{17.132}$$

3. The return flow, Q_2

$$Q_2 = \dot{x}_p (D^2 - d^2)\tfrac{\pi}{4} = \tfrac{Q_1}{D^2 \frac{\pi}{4}}(D^2 - d^2)\tfrac{\pi}{4}, \tag{17.133}$$

$$= \tfrac{(D^2-d^2)\frac{\pi}{4}}{D^2 \frac{\pi}{4}} Q_1 = \phi Q_1. \tag{17.134}$$

4. The cylinder chamber pressures, p_1 and p_2.

$$Q_2 = C_d A_0 \sqrt{\tfrac{2}{\rho}(p_2 - p_T)}, \tag{17.135}$$

$$p_2 = \frac{Q_2^2}{\left(C_d A_0 \sqrt{\frac{2}{\rho}}\right)^2} + p_T, \tag{17.136}$$

$$0 = \eta_{hmC} \left(p_1 D^2 - p_2 (D^2 - d^2)\right) \tfrac{\pi}{4} - mg, \tag{17.137}$$

$$p_1 = \tfrac{p_2 (D^2 - d^2)}{D^2} + \tfrac{mg}{D^2 \frac{\pi}{4} \eta_{hmC}}. \tag{17.138}$$

5. The power deliver to the load.

$$\mathcal{P}_{Load} = \dot{x}_p mg. \tag{17.139}$$

6. The hydraulic power delivered from the pump.

$$\mathcal{P}_P = Q_1 p_1. \tag{17.140}$$

7. The required mechanical power to the pump.

$$\mathcal{P}_{mech} = \frac{D_P n (p_1 - p_T)}{60 \eta_{hmP}}. \tag{17.141}$$

The results using the above equations are:

$$Q_1 = \underline{\underline{27\,\tfrac{\text{L}}{\text{min}}}},\quad \dot{x}_p = \underline{\underline{0.358\,\tfrac{\text{m}}{\text{s}}}},\quad Q_2 = \underline{\underline{20.25\,\tfrac{\text{L}}{\text{min}}}}, \tag{17.142}$$

$$p_2 = \underline{\underline{10\ \text{bar}}},\quad p_1 = \underline{\underline{89.7\ \text{bar}}}, \tag{17.143}$$

$$\mathcal{P}_{\text{Load}} = \underline{\underline{3.51\,\text{kW}}},\quad \mathcal{P}_{\text{Hyd}} = \underline{\underline{4.04\ \text{kW}}},\quad \mathcal{P}_{\text{Mech}} = \underline{\underline{4.72\ \text{kW}}}. \tag{17.144}$$

oOo

17.4 Steady State Analysis

17.4.1 System 1—Raising a Differential Cylinder

Exercise 16.4.1 is used as an example in the note, therefore the results are seen in Sect. 12.2

17.4.2 System 1—Lowering a Differential Cylinder

1. The piston velocity is found by calculating the pump outflow, assuming the pressure relief valve to be closed and using flow continuity equations for control volume P (pump to valve) and B (valve to piston on rod side).

$$\begin{aligned} Q_{\text{P}} &= D_{\text{P}} n, \\ Q_{\text{P}} &= 25\,\frac{\text{cm}^3}{\text{rev}}\,\frac{1\text{L}}{1000\,\text{cm}^3}\,2000\,\frac{\text{rev}}{\text{min}} = 50\,\frac{\text{L}}{\text{min}}. \end{aligned} \tag{17.145}$$

Control volume P:

$$\begin{aligned} Q_{\text{P}} - Q_{\text{pr}} + Q_{\text{B}} &= 0, \\ Q_{\text{B}} = -Q_{\text{P}} &= \underline{\underline{-50\tfrac{\text{L}}{\text{min}}}}. \end{aligned} \tag{17.146}$$

Control volume B yields:

$$-Q_{\text{B}} = -A_{\text{B}}\dot{x}_{\text{p}}, \tag{17.147}$$

$$\dot{x}_{\text{p}} = \frac{-0.000833\,\frac{\text{m}^3}{\text{s}}}{(0.04^2\,\text{m}^2 - 0.025^2\,\text{m}^2)\frac{\pi}{4}} = \underline{\underline{-1.09\,\tfrac{\text{m}}{\text{s}}}}. \tag{17.148}$$

2. The flow from A to tank T is given from the continuity equation for chamber A:

$$Q_A = A_A \dot{x}_p, \tag{17.149}$$

$$Q_A = 0.04^2\,\text{m}^2 \tfrac{\pi}{4}(-1.09)\,\tfrac{\text{m}}{\text{s}} 60\,\tfrac{\text{s}}{\text{min}} 1000\,\tfrac{\text{L}}{\text{m}^3} = \underline{\underline{-82.05\,\tfrac{\text{L}}{\text{min}}}}. \tag{17.150}$$

3. The pump pressure is calculated from the assumption of $p_B = 5$ bar and the pressure drop through the valve path PB as:

$$Q_{PB} = C_d A_0 \sqrt{\tfrac{2}{\rho}(p_P - p_B)}, \tag{17.151}$$

$$(p_P - p_B) = \left(\frac{Q_{PB}}{C_d A_0 \sqrt{\frac{2}{\rho}}}\right)^2, \tag{17.152}$$

$$(p_P - p_B) = \left(\frac{-0.000833\,\frac{\text{m}^3}{\text{s}}}{0.7 \cdot 10 \cdot 10^{-6}\,\text{m}^2 \sqrt{\frac{2}{900\,\frac{\text{kg}}{\text{m}^3}}}}\right)^2 = 19.68 \cdot 10^5\,\text{Pa}, \tag{17.153}$$

$$p_P = \underline{\underline{24.687 \cdot 10^5\,\text{Pa}}}. \tag{17.154}$$

4. As the B chamber pressure is given as 5 bar in the exercise, the A chamber pressure is the one to calculate from Newton's Second Law by:

$$M\ddot{x}_p = p_A A_A - p_B A_B - Mg = 0, \tag{17.155}$$

$$p_A = \frac{p_B A_B + Mg}{A_A}, \tag{17.156}$$

$$p_A = \frac{5 \cdot 10^5\,\text{Pa}(0.04^2\,\text{m}^2 - 0.025^2\,\text{m}^2)\frac{\pi}{4} + 2000\,\text{kg}\,9.81\,\frac{\text{m}}{\text{s}}^2}{0.04^2\,\text{m}^2\frac{\pi}{4}} = \underline{\underline{159.18 \cdot 10^5\,\text{Pa}}}. \tag{17.157}$$

5. The required flow area in the directional valve from A to T may now be found, as the opening area that leads to a pressure drop of 159.18 bar as a flow of 82.05 $\frac{\text{L}}{\text{min}}$ is flowing through. This area is found using the orifice equation as:

$$Q_{AT} = C_d A_{AT} \sqrt{\tfrac{2}{\rho}(p_A - p_T)}, \tag{17.158}$$

$$A_{AT} = \frac{Q_{AT}}{C_d \sqrt{\frac{2}{\rho}(p_A - p_T)}}, \tag{17.159}$$

$$A_{AT} = \frac{0.0014\,\frac{\text{m}^3}{\text{s}}}{0.7\sqrt{\frac{2}{900\,\frac{\text{kg}}{\text{m}^3}}(159.18\,\text{Pa} - 0)}} - \underline{\underline{10.4 \cdot 10^{-6}\,\text{m}^2}}. \tag{17.160}$$

17.4.3 System 2—Flow Regeneration 1

1. Flows have been added in Fig. 17.3, note that other definitions may be chosen, yielding equally good results.

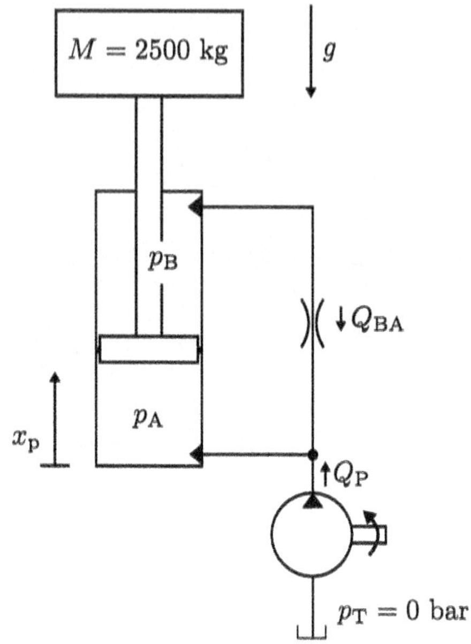

Fig. 17.3 Illustration for Exercises 16.4.3

2. The pump flow is given flow displacement volumetric efficiency and rotational velocity:

$$Q_P = D_P n \eta_{vP}, \tag{17.161}$$

$$Q_P = 30\,\frac{\text{cm}^3}{\text{rev}}\,\frac{1\,\text{L}}{1000\,\text{cm}^3}\,1000\,\frac{\text{rev}}{\text{min}}\,0.94 = \underline{\underline{28.2\,\frac{\text{L}}{\text{min}}}}. \tag{17.162}$$

3. The piston velocity is found by setting up flow continuity equations for chambers A and B and assuming steady state such that the pressure gradient is zero:

$$Q_P + Q_{BA} = \dot{x}_c A_A, \tag{17.163}$$

$$0 - Q_{BA} = -\dot{x}_c A_B \tag{17.164}$$

$$Q_P + \dot{x}_c A_B = \dot{x}_c A_A, \tag{17.165}$$

$$Q_P = \dot{x}_c (A_A - A_B), \tag{17.166}$$

$$\dot{x}_c = \frac{Q_P}{D^2\frac{\pi}{4} - (D^2 - d^2)\frac{\pi}{4}} = \frac{Q_P}{d^2\frac{\pi}{4}}\,\frac{0.00047\,\frac{\text{m}^3}{\text{s}}}{(0.040\,\text{m})^2\frac{\pi}{4}} = 0.37\,\frac{\text{m}}{\text{s}} \tag{17.167}$$

4. The pressure in the two cylinder chambers is found through force balance:

$$0 = (p_A A_A - p_B A_B)\eta_{hmC} - Mg, \tag{17.168}$$

$$p_B = p_A + \Delta p, \tag{17.169}$$

$$0 = (p_A A_A - (p_A + \Delta p)A_B)\eta_{hmC} - Mg, \tag{17.170}$$

$$p_A(A_A - A_B) = \frac{Mg}{\eta_{hmC}} + \Delta p A_B, \tag{17.171}$$

$$p_A = \frac{\frac{Mg}{\eta_{hmC}} + \Delta p A_B}{A_A - A_B}, \tag{17.172}$$

$$p_A = \frac{\frac{Mg}{\eta_{hmC}} + \Delta p A_B}{d^2 \frac{\pi}{4}} \tag{17.173}$$

The pressure difference from chamber B to A is found using the piston velocity and the orifice equation:

$$Q_1 = C_d A_O \sqrt{\frac{2}{\rho}\Delta p}, \tag{17.174}$$

$$\Delta p = \left(\frac{\dot{x}_p A_B}{C_d A_O \sqrt{\frac{2}{\rho}}}\right)^2, \tag{17.175}$$

$$\Delta p = \left(\frac{0.37\,\frac{m}{s}((0.050\,m)^2 - (0.040\,m)^2\frac{\pi}{4})}{0.610 \cdot 10^6\,m^2 \sqrt{\frac{2}{866\,\frac{kg}{m^3}}}}\right)^2 = 8.4 \cdot 10^5\,Pa \tag{17.176}$$

As such, the chamber pressure becomes:

$$p_A = \frac{\frac{Mg}{\eta_{hmC}} + \Delta p A_B}{d^2 \frac{\pi}{4}}, \tag{17.177}$$

$$p_A = \frac{\frac{2500\,kg 9.81\,\frac{m}{s^2}}{0.9} + 8.4 \cdot 10^5\,Pa(0.050\,m)^2 - (0.040\,m)^2\frac{\pi}{4}}{(0.040\,m^2\frac{\pi}{4}} = \underline{\underline{221.6\,bar}}, \tag{17.178}$$

$$p_B = p_A + \Delta p = 221.6\,bar + 8.4\,bar = \underline{\underline{230\,bar}} \tag{17.179}$$

5. The pressure will rise until the leakage flow in the pump equals the theoretical pump flow or something breaks and fluid spills out.

oOo

17.4.4 System 3—Flow Regeneration 2

(a) The flow P, 1, 2 and 3 have been added into the diagram, see Fig. 17.4.
(b) The pump flow is calculated from the pump displacement, pump velocity and volumetric efficiency:

$$Q_P = D_P n \eta_{vP}, \tag{17.180}$$

$$Q_P = 30\,\frac{cm^3}{rev}\frac{1\,L}{1000cm^3}1500\,\frac{rev}{min}0.9 = \underline{\underline{40.5\,\frac{L}{min}}}. \tag{17.181}$$

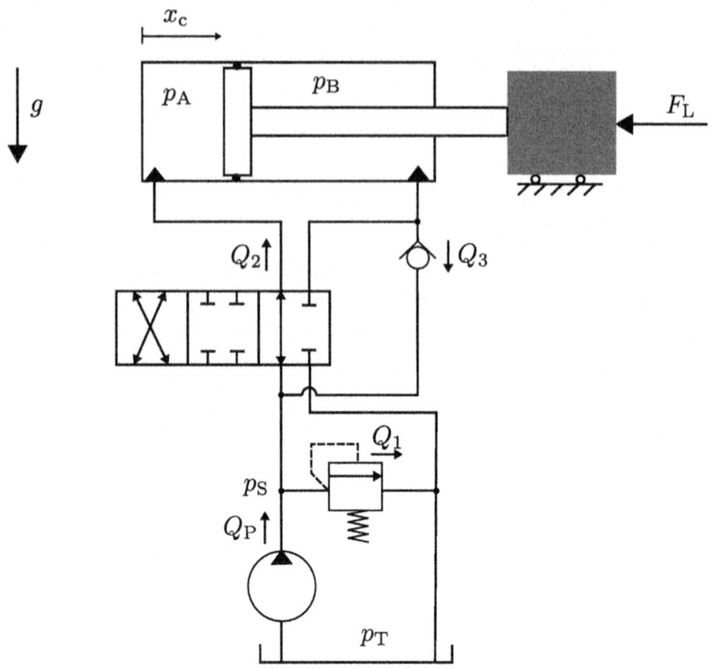

Fig. 17.4 Illustration for Exercises 16.4.4

(c) The piston velocity is found by setting up flow continuity equations for chambers A, B and the supply note and assuming steady state such that the pressure gradient is zero:

$$Q_P - Q_1 - Q_2 + Q_3 = 0, \tag{17.182}$$
$$Q_2 - 0 = \dot{x}_c A_A, \tag{17.183}$$
$$0 - Q_3 = -\dot{x}_c A_B. \tag{17.184}$$

Isolating Q_2 and Q_3 in the latter two equations and guessing that $Q_1 = 0$ (this should be checked latter).

$$Q_P - 0 - \dot{x}_c A_A + \dot{x}_c A_B = 0, \tag{17.185}$$
$$Q_P - 0 - \dot{x}_c(A_A - A_B) = 0, \tag{17.186}$$
$$\dot{x}_c = \frac{Q_P}{A_A - A_B}, \approx \underline{\underline{0.54\,\tfrac{\text{m}}{\text{s}}}}. \tag{17.187}$$

(d) As the check valve is assumed ideal the pressure p_B is equal to the supply pressure. The pressure in chamber A is lower than the supply line pressure due to a pressure drop across the "orifice" of the direction valve:

$$Q_2 = C_d A_0 \sqrt{\tfrac{2}{\rho}\Delta p}, \tag{17.188}$$

$$\Delta p = \left(\frac{Q_2}{C_d A_0 \sqrt{\frac{2}{\rho}}}\right)^2, \tag{17.189}$$

$$P_A = p_S - \Delta p. \tag{17.190}$$

Force equilibrium for the piston yields:

$$0 = p_A A_A - p_B A_B - F_L, \tag{17.191}$$

$$0 = (p_S - \Delta p)A_A - p_S A_B - F_L, \tag{17.192}$$

$$0 = p_S(A_A - A_B) - F_L - \Delta p A_A, \tag{17.193}$$

$$p_S = \frac{F_L + \Delta p A_A}{(A_A - A_B)} = \underline{\underline{110.1\,\text{bar}}}. \tag{17.194}$$

(e) The mechanical power delivered to the pump is given as:

$$\mathcal{P}_{\text{mech}} = \omega_P \tau_P = \omega_P \frac{D_P(p_S - p_T)}{\eta hm P}, \tag{17.195}$$

$$\mathcal{P}_{\text{mech}} = 157.08\,\text{rads}^{-1} \frac{4.7746 \cdot 10^{-6}\,\frac{\text{m}^3}{\text{rad}}(110.1 \cdot 10^5\,\text{Pa} - 0)}{0.85} = \underline{\underline{9.717\,\text{kW}}}. \tag{17.196}$$

(f) The power loss in the valve is calculated from the flow through it and the pressure drop across as:

$$\mathcal{P}_{\text{valve}} = (Q_P + \dot{x}_c A_B)(p_S - p_A), \tag{17.197}$$

$$\mathcal{P}_{\text{valve}} = 0.0011\,\frac{\text{m}^3}{\text{s}}\,19.55 \cdot 10^5\,\text{Pa} = \underline{\underline{2.063\,\text{kW}}}. \tag{17.198}$$

oOo

17.4.5 System 4—Motor Lifting the Load

The system flows are initially calculated from the pump to the tank.

(a) The full pump flow is assumed to go through the motor hence why velocity is:

$$Q_p = D_p n_p \eta_{pV} = 23\frac{\text{L}}{\text{min}}, \tag{17.199}$$

$$Q_m = Q_m, \tag{17.200}$$

$$n_m = \frac{Q_m \eta_{mV}}{D_m} = 230\,\frac{\text{rev}}{\text{min}}, \tag{17.201}$$

$$\dot{z} = \omega_m r = \underline{\underline{7.23\,\frac{\text{m}}{\text{s}}}}. \tag{17.202}$$

(b) The pressure in the system is calculated from the tank through the system and back to the pump.

$$p_4 = p_{\text{T}}, \tag{17.203}$$

$$p_3 = \frac{Q_{\text{BT}}^2}{C_{\text{d}}^2 A_{0,\text{BT}}^2 \sqrt{\frac{2}{\rho}}^2} + p_4 \approx \underline{\underline{7.00 \cdot 10^5\,\text{Pa}}}, \tag{17.204}$$

$$p_2 = \frac{\tau_{\text{m}}}{D_{\text{m}}\eta_{\text{mhm}}} + p_3 \approx \underline{\underline{212.46 \cdot 10^5\,\text{Pa}}}, \tag{17.205}$$

$$p_1 = \frac{Q_{\text{p}}^2}{C_{\text{d}}^2 A_{0,\text{PA}}^2 \sqrt{\frac{2}{\rho}}^2} + p_2 \approx \underline{\underline{218.46 \cdot 10^5\,\text{Pa}}}. \tag{17.206}$$

(c) The mechanical power required to drive the pump is found next as,

$$\mathcal{P} = \frac{D_{\text{p}}(p_1 - P_{\text{T}})n_{\text{p}}}{\eta_{mhp}} \approx \underline{\underline{10.067\,\text{kW}}}. \tag{17.207}$$

oOo

17.4.6 System 4—Motor Lowering the Load

In this exercise you will get various results depending on your system design.

oOo

17.5 System Modeling

17.5.1 System D1—Simple Pump Cylinder Drive

1. The equations used to calculate the cylinder piston velocity are: Newton's Second Law, ideal pump flow, laminar leakage and the flow continuity equation. As steady state is assumed, cylinder piston acceleration is zero.

$$p_{\text{A}} = \frac{F_{\text{L}} + M_{\text{L}}g + p_{\text{B}}A_{\text{B}}}{A_{\text{A}}} = \frac{F_{\text{L}} + 2500\,\text{kg}\,9.81\,\frac{\text{m}}{\text{s}^2}}{0.06^2\,\text{m}^2\,\frac{\pi}{4}}, \tag{17.208}$$

$$\dot{x}_{\text{c}} = \frac{Q_{\text{P}} - Q_{\text{le}}}{A_{\text{A}}}, \tag{17.209}$$

$$Q_{\text{P}} = D_{\text{P}}\omega_{\text{P}}, \tag{17.210}$$

$$Q_{\text{le}} = C_{\text{le}}\,p_{\text{A}}. \tag{17.211}$$

The ideal flow out of the pump is 30 $\frac{\text{L}}{\text{min}}$. The pressure in the A chamber is $p_{\text{A}} = [122.1\ 192.8\ 263.6]$ bar and this leads to a cylinder piston velocity of $\dot{x}_{\text{c}} = [0.173\ 0.171\ 0]\ \frac{\text{m}}{\text{s}}$. Note that the last velocity is zero as the requested pressure is above the setting of the "ideal" pressure relief valve.
2. A time domain model with pump velocity as input may be constructed out of one flow continuity equation, one Newton's Second Law, an ideal pump flow and a laminar leakage flow.

$$\ddot{x}_c = \frac{1}{M_L}(p_A A_A - M_L g - F_L), \tag{17.212}$$

$$\dot{p}_A = \frac{\beta}{V_{Ah} + A_A x_c}(D_P \omega_P - C_{le} p_A - \dot{x}_c A_A). \tag{17.213}$$

oOo

17.5.2 System D2—Hanging Mass

Initially we assumed the pressure p_l after the pump to be controlled by the pressure relief valve such that $p_l = p_{cr}$. We further assume the valve spool position to be a system input and as such disregard the valve dynamics (Table 17.2).

1. The time domain model chosen has the states, $[x_p\ \dot{x}_p\ p_A\ p_B]$ and consists of two continuity equations, a Newton's Second Law and two orifice equations:

$$\ddot{x}_p(t) = \frac{1}{M}(A_A p_A(t) - A_B p_B(t) - Mg + F_{fr}(t)), \tag{17.214}$$

$$\dot{p}_A(t) = \frac{\beta(p_A(t))}{V_A(t)}\left(Q_A(t) - \dot{x}_p(t) A_A\right), \tag{17.215}$$

$$\dot{p}_B(t) = \frac{\beta(p_B(t))}{V_B(t)}\left(\dot{x}_p(t) A_B - Q_B(t)\right). \tag{17.216}$$

As can be seen, no leakage is assumed in the cylinder. One now needs a model for volume, bulk modulus, friction and flows. The valve flows are modeled with the orifice equation derived from the data sheet to:

$$Q_A(t) = k_v|\bar{x}_v(t)| \begin{cases} \sqrt{|p_S(t) - p_A(t)|}\frac{p_S(t)-p_A(t)}{|p_S(t)-p_A(t)|}, & x_v(t) \geq 0 \\ \sqrt{|p_T(t) - p_A(t)|}\frac{p_T(t)-p_A(t)}{|p_T(t)-p_A(t)|}, & x_v(t) < 0 \end{cases}, \tag{17.217}$$

$$Q_B(t) = k_v|\bar{x}_v(t)| \begin{cases} \sqrt{|p_B(t) - p_T(t)|}\frac{p_B(t)-p_T(t)}{|p_B(t)-p_T(t)|}, & x_v(t) \geq 0 \\ \sqrt{|p_B(t) - p_S(t)|}\frac{p_B(t)-p_S(t)}{|p_B(t)-p_S(t)|}, & x_v(t) < 0 \end{cases}, \tag{17.218}$$

$$k_v = \frac{Q_{nom}}{\sqrt{p_{nom}}}. \tag{17.219}$$

The volumes are modelled as:

$$V_A(t) = V_{Hose.A} + A_A x_p(t), \tag{17.220}$$

$$V_B(t) = V_{Hose.B} + A_A(L - x_p(t)). \tag{17.221}$$

The friction model is based here on coulomb and viscous damping:

Table 17.2 Model data

Symbol	D	d	L	M	B	F_{c0}	Q_{nom}	p_{nom}
Value	0.040 m	0.025 m	0.400 m	120 kg	?	?	$10\,\frac{L}{min}$	35 bar

$$F_{\text{fr}}(t) = -F_{\text{c0}}\text{sign}(\dot{x}_\text{p}(t)) - B\dot{x}_\text{p}(t). \tag{17.222}$$

oOo

17.6 Analysis of Dynamics Systems—Solutions

17.6.1 Pilot Camber I

(a) The flow into the chamber is modelled with the orifice equation:

$$Q_{\text{Sc}} = C_\text{d}A_{0.1}\sqrt{\frac{2}{\rho}|p_\text{S} - p_\text{c}|}\frac{p_\text{S} - p_\text{c}}{|p_\text{S} - p_\text{c}|}. \tag{17.223}$$

(b) The pressure gradient is modelled with the continuity equation:

$$\dot{p}_\text{c} = \frac{\beta}{V_\text{c}}Q_{\text{Sc}} \tag{17.224}$$

(c) Assuming that bulk modulus is constant yields the continuity equation to be linear therefore linearisation with first order Taylor expansion is used only on the orifice equation:

$$\tilde{Q}_{\text{Sc}} = Q_{\text{Sc0}} + \frac{\partial Q_{\text{Sc}}}{\partial p_\text{S}}\bigg|_0 (p_\text{S} - p_{\text{S0}}) + \frac{\partial Q_{\text{Sc}}}{\partial p_\text{c}}\bigg|_0 (p_\text{c} - p_{\text{c0}}), \tag{17.225}$$

$$\Delta Q_{\text{Sc}} = \frac{\partial Q_{\text{Sc}}}{\partial p_\text{S}}\bigg|_0 \Delta p_\text{S} + \frac{\partial Q_{\text{Sc}}}{\partial p_\text{c}}\bigg|_0 \Delta p_\text{c}. \tag{17.226}$$

The partial derivatives are:

$$\frac{\partial Q_{\text{Sc}}}{\partial p_\text{S}}\bigg|_0 = \frac{C_\text{d}A_{0.1}\frac{2}{\rho}}{2\sqrt{\frac{2}{\rho}|p_{\text{S0}}-p_{\text{c0}}|}}, \tag{17.227}$$

$$\frac{\partial Q_{\text{Sc}}}{\partial p_\text{c}}\bigg|_0 = -\frac{C_\text{d}A_{0.1}\frac{2}{\rho}}{2\sqrt{\frac{2}{\rho}|p_{\text{S0}}-p_{\text{c0}}|}}, \tag{17.228}$$

$$k_{\text{qp}} = \frac{\partial Q_{\text{Sc}}}{\partial p_\text{S}}\bigg|_0 = \frac{C_\text{d}A_{0.1}\frac{2}{\rho}}{2\sqrt{\frac{2}{\rho}|p_{\text{S0}}-p_{\text{c0}}|}}, \tag{17.229}$$

The linear flow model is such that:

$$\Delta Q_{\text{Sc}} = k_{\text{qp}}\Delta p_\text{S} - k_{\text{qp}}\Delta p_\text{c}. \tag{17.230}$$

$$\Delta \dot{p}_\text{c} = \frac{\beta}{V_\text{c}}\Delta Q_{\text{Sc}}. \tag{17.231}$$

Laplace transforming yields:

$$Q_{Sc}(s) = k_{qp}P_S(s) - k_{qp}P_c(s), \tag{17.232}$$

$$P_c(s)s = \frac{\beta}{V_c}Q_{Sc}(s), \tag{17.233}$$

$$P_c(s)s = \frac{\beta}{V_c}(k_{qp}P_S(s) - k_{qp}P_c(s)). \tag{17.234}$$

(d) Setting up the transfer function:

$$P_c(s)\left(\frac{V_c}{\beta}s + k_{qp}P_c(s)\right) = k_{qp}P_S(s), \tag{17.235}$$

$$\frac{P_c(s)}{P_S(s)} = \frac{k_{qp}}{\frac{V_c}{\beta}s + k_{qp}}. \tag{17.236}$$

(e) The DC is 1. This is expected as fluid will flow between supply and the chamber until the two pressures are equal; hence as time passes the chamber pressure goes towards the supply pressure.

oOo

17.6.2 Pilot Camber II

(a) The flow into the chamber is modelled with the orifice equation:

$$Q_{Sc} = C_d A_{0.1}\sqrt{\frac{2}{\rho}|p_S - p_c|}\frac{p_S - p_c}{|p_S - p_c|}. \tag{17.237}$$

(b) The flow out of the chamber is modelled with the orifice equation:

$$Q_{cT} = C_d A_{0.2}\sqrt{\frac{2}{\rho}|p_c - p_T|}\frac{p_c - p_T}{|p_c - p_T|}. \tag{17.238}$$

(c) The pressure gradient is modelled with the continuity equation:

$$\dot{p}_c = \frac{\beta}{V_c}\left(Q_{Sc} - Q_{cT}\right). \tag{17.239}$$

(d) Firstly the orifice equations are linearised and next Laplace transformed:

$$\Delta Q_{Sc} = k_{qp.1}\Delta p_S - k_{qp.1}\Delta p_c, \tag{17.240}$$

$$\Delta Q_{cT} = k_{qp.2}\Delta p_c - k_{qp.2}\Delta p_T, \tag{17.241}$$

$$k_{qp.1} = \left.\frac{\partial Q_{Sc}}{\partial p_S}\right|_0 = \frac{C_d A_{0.1}\frac{2}{\rho}}{2\sqrt{\frac{2}{\rho}|p_{S0} - p_{c0}|}}, \tag{17.242}$$

$$k_{qp.2} = \left.\frac{\partial Q_{cT}}{\partial p_c}\right|_0 = \frac{C_d A_{0.2}\frac{2}{\rho}}{2\sqrt{\frac{2}{\rho}|p_{c0} - p_{T0}|}}. \tag{17.243}$$

Laplace transforming:

$$Q_{Sc}(s) = k_{qp.1} P_S(s) - k_{qp.1} P_c(s), \tag{17.244}$$
$$Q_{cT}(s) = k_{qp.2} P_c(s) - k_{qp.2} P_T(s), \tag{17.245}$$
$$P_c(s) \tfrac{V_c}{\beta} s = (Q_{Sc}(s) - Q_{cT}(s)), \tag{17.246}$$
$$P_c(s) \tfrac{V_c}{\beta} s = \left(k_{qp.1} P_S(s) - k_{qp.1} P_c(s) - k_{qp.2} P_c(s) + k_{qp.2} P_T(s)\right). \tag{17.247}$$

(e) The transfer function:

$$P_c(s)\left(\tfrac{V_c}{\beta} s + k_{qp.1} + k_{qp.2}\right) = k_{qp.1} P_S(s), \tag{17.248}$$
$$\tfrac{P_c(s)}{P_S(s)} = \tfrac{k_{qp.1}}{\frac{V_c}{\beta} s + k_{qp.1} + k_{qp.2}}. \tag{17.249}$$

(f) The DC gain is:

$$K_{DC} = \frac{k_{qp.1}}{k_{qp.1} + k_{qp.2}}. \tag{17.250}$$

The DC gain is seen to be given by the ratio between the two orifices.

oOo

17.6.3 Spring Loaded Accumulator

1. Flow through the two orifices is modelled with the orifice equation as:

$$Q_S = C_d A_{dS} x_v \sqrt{\tfrac{2}{\rho} |p_S - p_c|} \operatorname{sign}(p_S - p_c), \tag{17.251}$$
$$Q_T = C_d A_{dT} \sqrt{\tfrac{2}{\rho} |p_c - p_T|} \operatorname{sign}(p_c - p_T). \tag{17.252}$$

With $x_v = 0.25$ the orifice areas for the supply and tank valve are equal, which with steady flow yields the pressure drop across the two valves to be equal.

$$(p_S - p_c) = (p_c - p_T) = \tfrac{p_S - p_T}{2} = 50\text{bar}. \tag{17.253}$$

2. With $x_v = 0.25$ pressure in the accumulator is 50 bar. The piston position may be calculated with Newton's Second Law assuming force balance:

$$0 = p_c A - Mg - k_s x_p, \tag{17.254}$$
$$x_p = \tfrac{p_c A - Mg}{k_s} = \frac{50 \cdot 10^5 \text{Pa} 0.01^2\,\text{m}^2 \frac{\pi}{4} - 10\,\text{kg} 9.81\,\frac{\text{m}}{\text{s}^2}}{10000\,\frac{\text{N}}{\text{m}}} = 0.0295\ \text{m}. \tag{17.255}$$

3. Time domain dynamic equations are:

$$\dot{p}_c = \frac{\beta}{V_c(x_s)}\left(Q_S - Q_T - \dot{x}_p A\right), \tag{17.256}$$
$$\ddot{x}_p = \frac{1}{M}\left(p_c A - Mg - B\dot{x}_p - k_s x_p\right), \tag{17.257}$$
$$Q_S = C_d A_{dS} x_v \sqrt{\frac{2}{\rho}|p_S - p_c|}\text{sign}(p_S - p_c), \tag{17.258}$$
$$Q_T = C_d A_{dT} \sqrt{\frac{2}{\rho}|p_c - p_T|}\text{sign}(p_c - p_T). \tag{17.259}$$

4. In the linear model we assume the chamber volume to be constant in the linearisation point, $V_c(x_s) = V_{c0}$ and $\dot{x}_p A = 0$. The orifice equation is linearised with first order Taylor approximation and the linear model is made in change variables and the supply and tank pressure is assumed constant, hence why $\Delta p_S = \Delta p_T = 0$:

$$\Delta Q_S = k_{qSx}\Delta x_v - k_{qSp}\Delta p_c \tag{17.260}$$
$$\Delta Q_T = k_{qTp}\Delta p_c, \tag{17.261}$$
$$k_{qSx} = \left.\frac{\partial Q_S}{\partial x_v}\right|_0 = C_d A_{dS}\sqrt{\frac{2}{\rho}(p_{S0} - p_{c0})} \tag{17.262}$$
$$k_{qSp} = \left.\frac{\partial Q_S}{\partial p_S}\right|_0 = \frac{C_d A_{dS} x_{v0} \frac{2}{\rho}}{2\sqrt{\frac{2}{\rho}(p_{S0} - p_{c0})}}, \tag{17.263}$$
$$k_{qTp} = \left.\frac{\partial Q_T}{\partial p_c}\right|_0 = \frac{C_d A_{dT} \frac{2}{\rho}}{2\sqrt{\frac{2}{\rho}(p_{c0} - p_{T0})}}, \tag{17.264}$$

Laplace transforming and setting up the transfer function:

$$P_c(s)\frac{V_{c0}}{\beta} = \left(k_{qSx}X_v(s) - k_{qSp}P_c(s) - k_{qTp}P_c(s)\right), \tag{17.265}$$
$$P_c(s)\left(\frac{V_{c0}}{\beta}s + k_{qSp} + k_{qTp}\right) = k_{qSx}X_v(s), \tag{17.266}$$
$$P_c(s) = \frac{k_{qSx}X_v(s)}{\frac{V_{c0}}{\beta}s + k_{qSp} + k_{qTp}}, \tag{17.267}$$
$$X_p(s)\left(Ms^2 + Bs + k_s\right) = P_c(s)A, \tag{17.268}$$
$$X_p(s)\left(Ms^2 + Bs + k_s\right) = \frac{k_{qSx}X_v(s)}{\frac{V_{c0}}{\beta}s + k_{qSp} + k_{qTp}}A, \tag{17.269}$$
$$\frac{X_p(s)}{X_v(s)} = \frac{k_{qSx}A}{\frac{V_{c0}}{\beta}s + k_{qSp} + k_{qTp}}\frac{1}{Ms^2 + Bs + k_s}. \tag{17.270}$$

5. DC-gain at linearisation point $x_{v0} = 0.25$ at which the pressure and position is 50 bar and 0.0295 m m respectively. At DC $s = 0$:

$$\frac{X_p(0)}{X_v(0)} = \frac{k_{qSx}}{k_{qSp} + k_{qTp}}\frac{A}{k_s} = 0.1571 \tag{17.271}$$

oOo

17.6.4 Analysis of System D1

This solution continues on from the model in Sect. 17.5.1. By assuming the A chamber volume and the bulk modulus constant to $V_A(x_{c0})$ and $\beta = 10000$ bar only Newton's Second Law is non-linear.

1. In deriving a transfer function, firstly a linear model is derived then Laplace transformed:

$$\Delta \ddot{x}_c = \tfrac{1}{M_L}\left(\Delta p_A A_A - \Delta F_L\right), \tag{17.272}$$

$$\Delta \dot{p}_A = \tfrac{\beta}{V_A(x_{c0})}\left(D_P \Delta\omega_P - C_{le}\Delta p_A - \Delta\dot{x}_c A_A\right). \tag{17.273}$$

Laplace transformation yields:

$$X_c(s) M_L s^2 = P_A(s) A_A - F_L(s), \tag{17.274}$$

$$P_A(s)\tfrac{V_A(x_{c0})}{\beta}s = D_P\Omega_P(s) - C_{le}P_A(s) - sX_c(s)A_A. \tag{17.275}$$

$C_{le} = 0$ and $F_L(s) = 0$, yields:

$$P_A(s) = \frac{D_P\Omega_P(s) - sX_c(s)A_A}{\frac{V_A(x_{c0})}{\beta}s}, \tag{17.276}$$

$$X_c(s)M_L s^2 = \frac{D_P\Omega_P(s) - sX_c(s)A_A}{\frac{V_A(x_{c0})}{\beta}s}A_A, \tag{17.277}$$

$$sX_c(s)\left(\tfrac{M_L V_A(x_{c0})}{\beta}s^2 + A_A^2\right) = A_A D_P \Omega_P(s), \tag{17.278}$$

$$\frac{sX_c(s)}{\Omega_P(s)} = \frac{A_A D_P}{\left(\frac{M_L V_A(x_{c0})}{\beta}s^2 + A_A^2\right)} = \frac{A_A D_P \frac{\beta}{M_L V_A(x_{c0})}}{\left(s^2 + \frac{\beta A_A^2}{M_L V_A(x_{c0})}\right)}. \tag{17.279}$$

2. The highest and lowest eigenfrequency of the system:

$$\omega_n = \sqrt{\frac{\beta A_A^2}{M_L V_A(x_{c0})}} \tag{17.280}$$

The eigenfrequency is seen to be dependent only on the cylinder position (in regard to linearisation point); the highest and lowest are as such found for $x_{c0} = 0$ m and $x_{c0} = 0.5$ m respectively and are 178.82 $\frac{\text{rad}}{\text{s}}$ and 45.96 $\frac{\text{rad}}{\text{s}}$ respectively.

3. The DC-gain from pump velocity to cylinder piston velocity is:

$$\frac{sX_c(0)}{\Omega_P(0)} = \frac{A_A D_P}{A_A^2} = \frac{D_P}{A_A}. \tag{17.281}$$

oOo

17.6.5 Analysis of System D2

oOo

17.6.6 System D3—Hanging Mass

1. The model is linearised for $x_{\mathrm{v}}(t) \geq 0$. In Newton's Second Law, linearisation entails that the coulomb and gravitational forces' terms disappear. With bulk modulus and the chamber volumes constant, the continuity equations describing the pressure gradient are linear:

$$\Delta\ddot{x}_{\mathrm{p}}(t) = \tfrac{1}{M}\left(A_{\mathrm{A}}\Delta p_{\mathrm{A}}(t) - A_{\mathrm{B}}\Delta p_{\mathrm{B}}(t) - B\Delta\dot{x}_{\mathrm{p}}(t)\right), \tag{17.282}$$

$$\Delta\dot{p}_{\mathrm{A}}(t) = \tfrac{\beta(p_{\mathrm{A0}})}{V_{\mathrm{A0}}}\left(\Delta Q_{\mathrm{A}}(t) - \Delta\dot{x}_{\mathrm{p}}(t)A_{\mathrm{A}}\right), \tag{17.283}$$

$$\Delta\dot{p}_{\mathrm{B}}(t) = \tfrac{\beta(p_{\mathrm{B0}})}{V_{\mathrm{B0}}}\left(\Delta\dot{x}_{\mathrm{p}}(t)A_{\mathrm{B}} - \Delta Q_{\mathrm{B}}(t)\right). \tag{17.284}$$

Note that the linear model is in change variables, hence the Δ notation. Linearisation point values are indicated by including a 0 in the subscript. The valve flows given with orifice equations are linearised employing first order Taylor approximation:

$$\Delta Q_{\mathrm{A}}(t) = k_{\mathrm{qAx}}\Delta\bar{x}_{\mathrm{v}}(t) + k_{\mathrm{qAp}}\Delta p_{\mathrm{S}}(t) - k_{\mathrm{qAp}}\Delta p_{\mathrm{A}}(t), \tag{17.285}$$

$$\Delta Q_{\mathrm{B}}(t) = k_{\mathrm{qBx}}\Delta\bar{x}_{\mathrm{v}}(t) + k_{\mathrm{qBp}}\Delta p_{\mathrm{B}}(t) - k_{\mathrm{qBp}}\Delta p_{\mathrm{T}}(t). \tag{17.286}$$

The linearisation constants/partial derivatives are:

$$k_{\mathrm{qAx}} = \left.\frac{\partial Q_{\mathrm{A}}}{\partial x_{\mathrm{v}}}\right|_0 = k_{\mathrm{v}}\sqrt{p_{\mathrm{S0}} - p_{\mathrm{A0}}}, \tag{17.287}$$

$$k_{\mathrm{qAp}} = \left.\frac{\partial Q_{\mathrm{A}}}{\partial p_{\mathrm{S}}}\right|_0 = \frac{k_{\mathrm{v}}\bar{x}_{\mathrm{v0}}}{2\sqrt{p_{\mathrm{S0}} - p_{\mathrm{A0}}}}, \tag{17.288}$$

$$k_{\mathrm{qBx}} = \left.\frac{\partial Q_{\mathrm{B}}}{\partial x_{\mathrm{v}}}\right|_0 = k_{\mathrm{v}}\sqrt{p_{\mathrm{B0}} - p_{\mathrm{T0}}}, \tag{17.289}$$

$$k_{\mathrm{qBp}} = \left.\frac{\partial Q_{\mathrm{B}}}{\partial p_{\mathrm{B}}}\right|_0 = \frac{k_{\mathrm{v}}\bar{x}_{\mathrm{v0}}}{2\sqrt{p_{\mathrm{B0}} - p_{\mathrm{T0}}}}. \tag{17.290}$$

2. For the system to become an LTI system after linearisation, the linearisation point, $\mathbf{x}_0$, must be an equilibrium point such that $\mathbf{0} = \mathbf{f}(\mathbf{x}_0)$. This means that if the system is "placed" at $\mathbf{x}_0 = [x_{\mathrm{p0}}\ \dot{x}_{\mathrm{p0}}\ p_{\mathrm{A0}}\ p_{\mathrm{B0}}]$ corresponding to x_{v0} the states do not change if the input is not changed. With no leakage in the system a proper choice of valve position is $x_{\mathrm{v}} = 0$ yielding zero cylinder piston velocity. The A and B pressures must obey:

$$0 = A_A p_{A0} - A_B p_{B0} + Mg, \tag{17.291}$$

$$p_{B0} = \frac{A_A p_{A0} + Mg}{A_B}. \tag{17.292}$$

The A pressure and the cylinder position may be chosen "freely".

3. State Space form

$$\dot{\mathbf{x}} = \mathbf{A}\mathbf{x} + \mathbf{B}\mathbf{u}, \tag{17.293}$$

$$\mathbf{y} = \mathbf{C}\mathbf{x} + \mathbf{D}\mathbf{u} \tag{17.294}$$

$$\dot{\mathbf{x}} = \begin{bmatrix} 0 & 1 & 0 & 0 \\ 0 & \frac{-B}{M} & \frac{A_A}{M} & \frac{-A_B}{M} \\ 0 & \frac{-\beta_{A0} A_A}{V_{A0}} & \frac{-\beta_{A0} k_{qAp}}{V_{A0}} & 0 \\ 0 & \frac{\beta_{B0} A_B}{V_{B0}} & \frac{\beta_{B0} k_{qBp}}{V_{B0}} & 0 \end{bmatrix} \mathbf{x} + \begin{bmatrix} 0 & 0 & 0 \\ 0 & 0 & 0 \\ \frac{\beta_{A0} k_{qAx}}{V_{A0}} & \frac{\beta_{A0} k_{qAp}}{V_{A0}} & 0 \\ \frac{\beta_{B0} k_{qBx}}{V_{B0}} & 0 & \frac{\beta_{B0} k_{qBp}}{V_{B0}} \end{bmatrix} \mathbf{u}. \tag{17.295}$$

The state and input vectors are $\mathbf{x} = [x_p \; \dot{x}_p \; p_A \; p_B]$ and $\mathbf{u} = [x_v \; p_S \; p_T]$ respectively. Note that both the supply and tank pressures are included as inputs yielding a multiple input system; when these are set constant a single input system emerges. One may note that assuming the volumes constant at the linearisation point brings a system matrix with no values in the first column. In this way, the system matrix is independent of the cylinder piston position and this state may be omitted in the model.

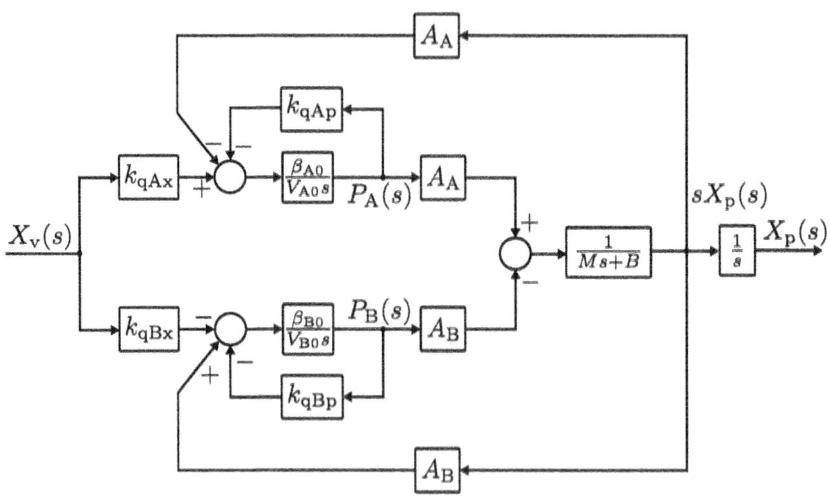

Fig. 17.5 Block diagram

$$\dot{\mathbf{z}} = \begin{bmatrix} \frac{-B}{M} & \frac{A_A}{M} & \frac{-A_B}{M} \\ \frac{-\beta_{A0}A_A}{V_{A0}} & \frac{-\beta_{A0}k_{qAp}}{V_{A0}} & 0 \\ \frac{\beta_{B0}A_B}{V_{B0}} & \frac{\beta_{B0}k_{qBp}}{V_{B0}} & 0 \end{bmatrix} \mathbf{z} + \begin{bmatrix} 0 & 0 & 0 \\ \frac{\beta_{A0}k_{qAx}}{V_{A0}} & \frac{\beta_{A0}k_{qAp}}{V_{A0}} & 0 \\ \frac{\beta_{B0}k_{qBx}}{V_{B0}} & 0 & \frac{\beta_{B0}k_{qBp}}{V_{B0}} \end{bmatrix} \mathbf{u}. \quad (17.296)$$

The state and input vectors are $\mathbf{z} = [\dot{x}_p \; p_A \; p_B]$ and $\mathbf{u} = [x_v \; p_S \; p_T]$ respectively.

4. Laplace transforming and assuming constant supply and tank pressure yields:

$$X_p(s)\left(Ms^2 + Bs\right) = A_A P_A(s) - A_B P_B(s), \quad (17.297)$$

$$P_A(s)\frac{V_{A0}}{\beta_{A0}}s = k_{qAx}X_v(s) - k_{qAp}P_A(s) - sX_p(s)A_A, \quad (17.298)$$

$$P_B(s)\frac{V_{B0}}{\beta_{B0}}s = sX_p(s)A_B - k_{qBx}X_v(s) - k_{qBp}P_B(s). \quad (17.299)$$

Setting the equations up in a block diagram yields Fig. 17.5. The figure shows a block diagram with valve spool position as input and cylinder piston position as output.

oOo

The manufacturer's authorised representative in the EU is Springer Nature Customer Service Centre GmbH, Europaplatz 3, 69115 Heidelberg, Germany. If you have any concerns regarding our products, please contact ProductSafety@springernature.com

Printed and bound by CPI Group (UK) Ltd, Croydon, CR0 4YY
30/07/2026
02182050-0002